W9-CCE-969

The Cross-Platform Prep Course

Welcome to the Cross-Platform Prep Course! McGraw-Hill Education's multi-platform course gives you a variety of tools to help you raise your test scores. Whether you're studying at home, in the library, or on-the-go, you can find practice content in the format you need—print, online, or mobile.

Print Book

This print book gives you the tools you need to ace the test. In its pages you'll find smart test-taking strategies, in-depth reviews of key topics, and ample practice questions and tests. See the Welcome section of your book for a step-by-step guide to its features.

Online Platform

The Cross-Platform Prep Course gives you additional study and practice content that you can access *anytime, anywhere*. You can create a personalized study plan based on your test date that sets daily goals to keep you on track. Integrated lessons provide important review of key topics. Practice questions, exams, and flashcards give you the practice you need to build test-taking confidence. The game center is filled with challenging games that allow you to practice your new skills in a fun and engaging way. And, you can even interact with other test-takers in the discussion section and gain valuable peer support.

Getting Started

To get started, open your account on the online platform:

Go to www.xplatform.mhprofessional.com

↓

Enter your access code, which you can find on the inside back cover of your book

↓

Provide your name and e-mail address to open your account and create a password

↓

Click "Start Studying" to enter the platform

It's as simple as that. You're ready to start studying online.

Your Personalized Study Plan

First, select your test date on the calendar, and you'll be on your way to creating your personalized study plan. Your study plan will help you stay organized and on track and will guide you through the course in the most efficient way. It is tailored to *your* schedule and features daily tasks that are broken down into manageable goals. You can adjust your test date at any time and your daily tasks will be reorganized into an updated plan.

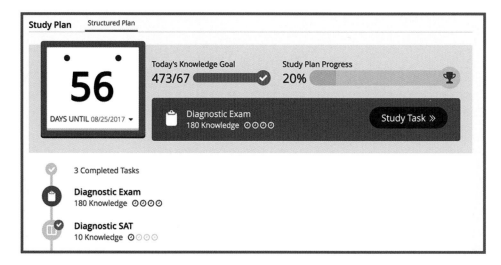

You can track your progress in real time on the Study Plan Dashboard. The "Today's Knowledge Goal" progress bar gives you up-to-the minute feedback on your daily goal. Fulfilling this every time you log on is the most efficient way to work through the entire course. You always get an instant view of where you stand in the entire course with the Study Plan Progress bar.

If you need to exit the program before completing a task, you can return to the Study Plan Dashboard at any time. Just click the Study Task icon and you can automatically pick up where you left off.

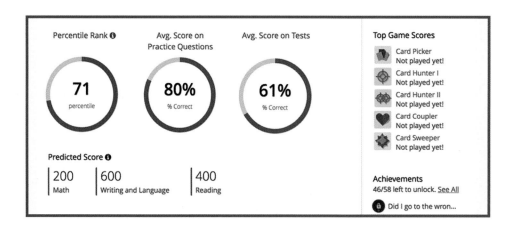

Practice Tests

One of the first tasks in your personalized study plan is to take the Diagnostic Test. At the end of the test, a detailed evaluation of your strengths and weaknesses shows the areas where you need the most focus. You can review your practice test results either by the question category to see broad trends or question-by-question for a more in-depth look.

The full-length tests are designed to simulate the real thing. Try to simulate actual testing conditions and be sure you set aside enough time to complete the full-length test. You'll learn to pace yourself so that you can work toward the best possible score on test day.

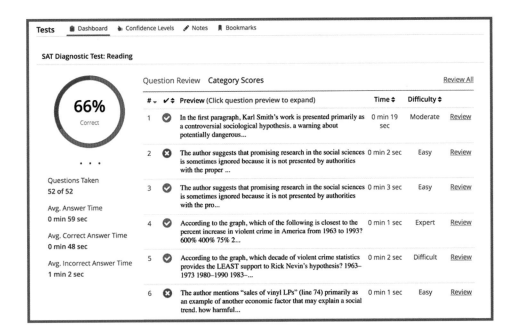

Lessons

The lessons in the online platform are divided into manageable pieces that let you build knowledge and confidence in a progressive way. They cover the full range of topics that you're likely to see on your test.

After you complete a lesson, mark your confidence level. (You must indicate a confidence level in order to count your progress and move on to the next task.) You can also filter the lessons by confidence levels to see the areas you have mastered and those that you might need to revisit.

> *Use the bookmark feature to easily refer back to a concept or leave a note to remember your thoughts or questions about a particular topic.*

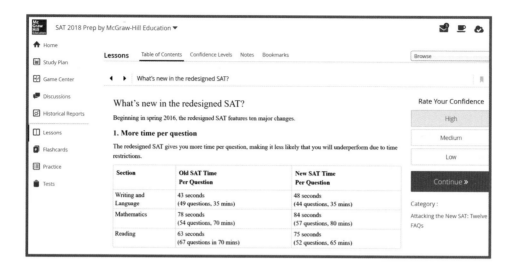

Practice Questions

All of the practice questions are reflective of actual exams and simulate the test-taking experience. The "Review Answer" button gives you immediate feedback on your answer. Each question includes a rationale that explains why the correct answer is right and the others are wrong. To explore any topic further, you can find detailed explanations by clicking the "Help me learn about this topic" link.

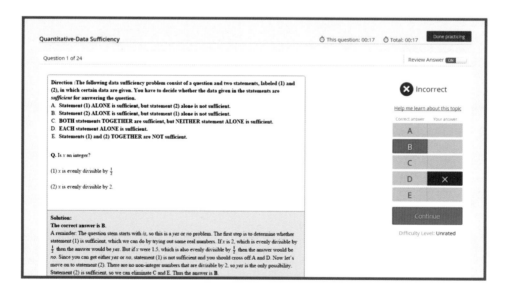

You can go to the Practice Dashboard to find an overview of your performance in the different categories and sub-categories.

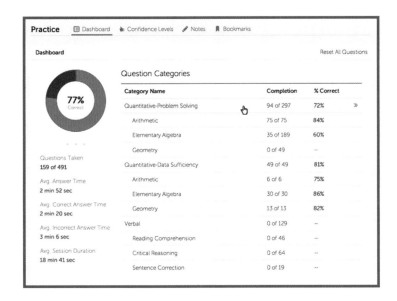

Dashboard

The dashboard is constantly updating to reflect your progress and performance. The Percentile Rank icon shows your position relative to all the other students enrolled in the course. You can also find information on your average scores in practice questions and exams.

A detailed overview of your strengths and weaknesses shows your proficiency in a category based on your answers and difficulty of the questions. By viewing your strengths and weaknesses, you can focus your study on areas where you need the most help.

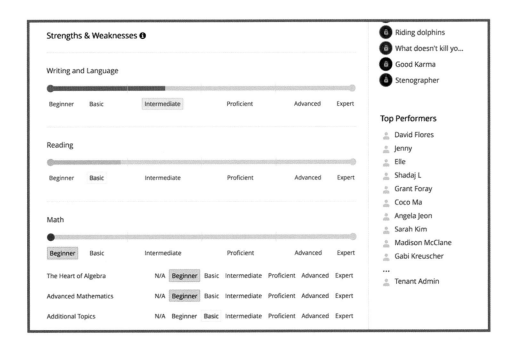

Flashcards

The hundreds of flashcards are perfect for learning key terms quickly, and the interactive format gives you immediate feedback. You can filter the cards by category and confidence level for a more organized approach. Or, you can shuffle them up for a more general challenge.

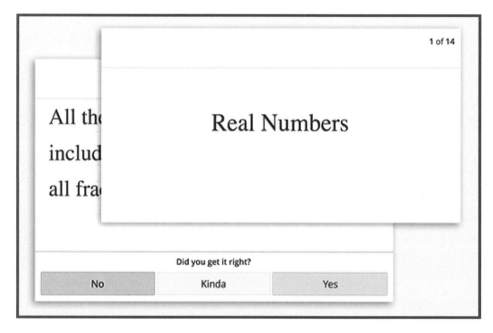

Another way to customize the flashcards is to create your own sets. You can either keep these private or share or them with the public. Subscribe to Community Sets to access sets from other students preparing for the same exam.

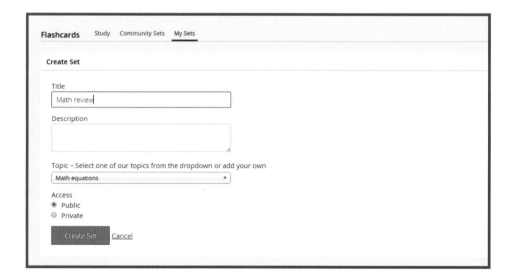

Game Center

Play a game in the Game Center to test your knowledge of key concepts in a challenging but fun environment. Increase the difficulty level and complete the games quickly to build your highest score. Be sure to check the leaderboard to see who's on top!

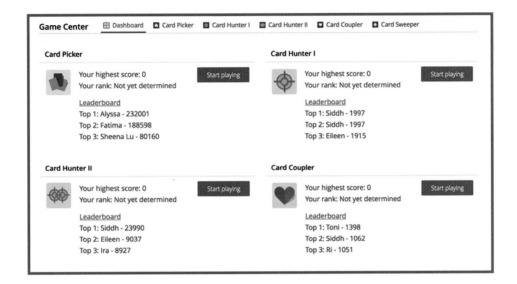

Social Community

Interact with other students who are preparing for the same test. Start a discussion, reply to a post, or even upload files to share. You can search the archives for common topics or start your own private discussion with friends.

Mobile App

The companion mobile app lets you toggle between the online platform and your mobile device without missing a beat. Whether you access the course online or from your smartphone or tablet, you'll pick up exactly where you left off.

Go to the iTunes or Google Play stores and search "McGraw-Hill Education Cross-Platform App" to download the companion iOS or Android app. Enter your e-mail address and the same password you created for the online platform to open your account.

Now, let's get started!

5 STEPS TO A 5™

AP Biology
2019

Mark Anestis • Kellie Ploeger Cox, PhD

Mc Graw Hill Education

New York Chicago San Francisco Athens London Madrid
Mexico City Milan New Delhi Singapore Sydney Toronto

1 2 3 4 5 6 7 8 9   LHS   23 22 21 20 19 18
1 2 3 4 5 6 7 8 9   LHS   23 22 21 20 19 18 (Elite Student Edition)

ISBN 978-1-260-12281-7
MHID 1-260-12281-6

e-ISBN 978-1-260-12282-4
e-MHID 1-260-12282-4

ISBN 978-1-260-12283-1 (Elite Student Edition)
MHID 1-260-12283-2

e-ISBN 978-1-260-12284-8 (Elite Student Edition)
e-MHID 1-260-12284-0

Trademarks: McGraw-Hill Education, the McGraw-Hill Education logo, *5 Steps to a 5*, and related trade dress are trademarks or registered trademarks of McGraw-Hill Education and/or its affiliates in the United States and other countries and may not be used without written permission. All other trademarks are the property of their respective owners. McGraw-Hill Education is not associated with any product or vendor mentioned in this book.

AP, Advanced Placement Program, and *College Board* are registered trademarks of the College Board, which was not involved in the production of, and does not endorse, this product.

The series editor was Grace Freedson, and the project editor was Del Franz.

Series design by Jane Tenenbaum.

McGraw-Hill Education products are available at special quantity discounts to use as premiums and sales promotions, or for use in corporate training programs. To contact a representative, please visit the Contact Us pages at www.mhprofessional.com.

CONTENTS

STEP 5 Build Your Test-Taking Confidence

Hello, and welcome to the new edition of the AP Biology review book that promises to be the most fun you have ever had!!!! Well, OK. . . . It will not be the most fun you have ever had . . . but maybe you will enjoy yourself a little bit. If you let yourself, you may at least learn a lot from this book. It contains the major concepts and ideas to which you were exposed over the past year in your AP Biology classroom, written in a manner that, we hope, will be pleasing to your eyes and your brain.

Many books on the market contain the same information that you will find in this book. However, we have approached the material a bit differently. We have tried to make this book as conversational and understandable as possible. We have had to review for countless standardized tests and cannot think of anything more annoying than a review book that is a total snoozer. In fact, we had this book "snooze-tested" by more than five thousand students, and the average reader could go 84 pages before falling asleep. This is better than the "other" review books whose average snooze time fell within the range of 14–43 pages. OK, we made up those statistics . . . but we promise that this book will not put you to sleep.

While preparing this book, we spoke to 154,076 students who had taken the AP exam and asked them how they prepared for the test. They indicated which study techniques were most helpful to them and which topics in this book they considered *vital* to success on this test. Throughout the book there are notes in the margins with these students' comments and tips. Pay heed to these comments because these folks know what they are talking about. They have taken this test and may have advice that will be useful to you.

We are not going to mislead you into thinking that you do not need to study to do well on this exam. You will actually need to prepare quite a bit. But this book will walk you through the process in as painless a way as possible. Use the study questions at the end of each chapter in Step 4 to practice applying the material you just read. Use the study tips listed in Step 3 to help you remember the material you need to know. Take the two practice tests in Step 5 as the actual exam approaches to see how well the information is sinking in.

Well, it's time to stop gabbing and start studying. Begin by setting up your study program in Step 1 of this book. Take the diagnostic test in Step 2 and look through the answers and explanations to see where you stand before you dive into the review process. Then look through the hints and strategies in Step 3, which may help you finally digest all the information that comes at you in Step 4. Then, we suggest that you kick back, relax, grab yourself a comfortable seat, and dig in. There is a lot to learn before the exam. Happy reading!

ACKNOWLEDGMENTS

This project would not have been completed without the assistance of many dear friends and relatives. To my wife, Stephanie, your countless hours of reading, rereading, and reading once again were of amazing value. Thank you so much for putting in so much time and energy to my cause. You have helped make this book what it is. To my parents and brothers who likewise contributed by reading a few chapters when I needed a second opinion, I thank you. I would like to thank Chris Black for helping me edit and clarify a few of the chapters. I would like to thank Don Reis, whose editing comments have strengthened both the content and the flow of this work. Finally, a big thank-you to all the students and teachers who gave me their input and thoughts on what they believed to be important for this exam. They have made this book that much stronger. Thank you all.

—Mark Anestis

ABOUT THE AUTHORS

MARK ANESTIS was born in Pittsburgh, Pennsylvania, and has lived in Connecticut since the age of six. He graduated from Weston High School in Weston, Connecticut, in 1993 and attended Yale University. While taking science courses in preparation for medical school, he earned a bachelor's degree cum laude in economics. He attended the University of Connecticut School of Medicine for two years and passed the step 1 boards, then chose to redirect his energy toward educating students in a one-on-one environment. He is the founder and director of The Learning Edge, a tutoring company based in Hamden, Connecticut (www.thelearningedge.net). Since January 2000, he has been tutoring high school students in math, the sciences, and standardized test preparation (including the SAT, ACT, and SAT Subject Tests). In addition to this review book, he has coauthored *McGraw-Hill's SAT, McGraw-Hill's PSAT/NMSQT*, and *McGraw-Hill's 12 SAT Practice Tests and PSAT*. He lives with his wife and sons in Hamden, Connecticut.

The author also created the PERFECT SCORE app (perfect-score.com), which allows students to prepare for the SAT while competing against other students. The app is available on the App Store.

KELLIE PLOEGER COX is originally from Spokane, Washington, and earned her PhD in Microbiology from the University of Idaho. She now lives in Connecticut, where she has been a member of the science department at Hopkins School since 1999. Kellie teaches Introductory Biology, AP Biology, and Contemporary Issues in Biology, and loves all things nerdy. Her favorite science experiments are her two daughters, Raina Faith and Kenley Piper.

INTRODUCTION: THE FIVE-STEP PROGRAM

Welcome!

If you focus on the beginning, the rest will fall into place. When you purchase this book and decide to work your way through it, you are beginning your journey to the Advanced Placement (AP) Biology exam. We will be with you every step of the way.

Why This Book?

We believe that this book has something unique to offer you. We have spoken with many AP Biology teachers and students and have been fortunate to learn quite a bit. Teachers helped us understand which topics are most important for class, and students described what they wanted from a test-prep book. Therefore, the contents of this book reflect genuine student concerns and needs. This is a student-oriented book. We did not attempt to impress you with arrogant language, mislead you with inaccurate information and tasks, or lull you into a false sense of confidence through ingenious shortcuts. We have not put information into this book simply because it is included in other review books. We recognize the fact that there is only so much that one individual can learn for an exam. Believe us, we have taken our fair share of these tests—we know how much work they can be. This book represents a realistic approach to studying for the AP exam. We have included very little heavy technical detail in this book. (There *is* some . . . we had to . . . but there is not very much.)

Think of this text as a resource and guide to accompany you on your AP Biology journey throughout the year. This book is designed to serve many purposes. It should:

- Clarify requirements for the AP Biology exam
- Provide you with test practice
- Help you pace yourself
- Function as a wonderful paperweight when the exam is completed
- Make you aware of the Five Steps to Mastering the AP Biology Exam

Organization of the Book

We know that your primary concern is to learn about the AP Biology exam. We start by introducing the five-step plan. We then give an overview of the exam in general. We follow that up with three different approaches to exam preparation and then move on to describe some tips and suggestions for how to approach the various sections of the exam. The Diagnostic Exam should give you an idea of where you stand before you begin your preparations. We recommend that you spend 45 minutes on this practice exam.

The volume of material covered in AP Biology is quite intimidating. Step 4 of this book provides a comprehensive review of all the major sections you may or may not have covered in the classroom. Not every AP Biology class in the country will get through the same amount of material. This book should help you fill any gaps in your understanding of the course work.

Step 5 of this book is the practice exam section. Here is where you put your skills to the test. The multiple-choice questions provide practice with the types of questions you may encounter on the AP exam. Keep in mind that they are *not* exact questions taken directly from past exams. Rather, they are designed to focus you on the key topics that often appear on the actual AP Biology exam. When you answer a question we've written in this book, do not think to yourself, "OK . . . that's a past exam question." Instead, you should think to yourself, "OK, the authors thought that was important, so I should remember this fact. It may show up in some form on the real exam." The essay questions are designed to cover the techniques and terms required by the AP exam. After taking each exam, you can check yourself against the explanations of every multiple-choice question and the grading guidelines for the essays.

The material at the end of the book is also important. It contains a bibliography of sources that may be helpful to you, a list of websites related to the AP Biology exam, and a glossary of the key terms discussed in this book.

Introducing the Five-Step Program

The five-step program is designed to provide you with the skills and strategies vital to the exam and the practice that can help lead you to that perfect "Holy Grail" score of 5.

Step 1: Set Up Your Study Program

Step 1 leads you through a brief process to help determine which type of exam preparation you want to commit yourself to:

1. Full-year prep: September through May
2. One-semester prep: January through May
3. Six-week prep: the six weeks prior to the exam

Step 2: Determine Your Test Readiness

Step 2 consists of a diagnostic exam, which will give you an idea of what you already know and what you need to learn between now and the exam. Take the test, which is broken down by topic, look over the detailed explanations, and start learning!

Step 3: Develop Strategies for Success

Step 3 gives you strategy advice for the AP Biology exam. It teaches you about the multiple-choice questions and the free-response questions you will face on exam day.

Step 4: Review the Knowledge You Need to Score High

Step 4 is a big one. This is the comprehensive review of all the topics on the AP exam. You've probably been in an AP Bio class all year, and you've likely spent hours upon hours reading through the AP Biology textbooks. These review chapters are appropriate both for quick skimming (to remind yourself of salient points that may have slipped your mind) and for in-depth study (to teach yourself broader concepts that may be new to you.)

Step 5: Build Your Test-Taking Confidence

Ahhhh, the full-length practice tests—oh, the joy! This book has three of them. If you purchased the cross-platform version of this book, you will find three more practice tests online. One of the most effective ways to improve as you prepare for any exam is to take as many practice tests as you can. Sit down and take these tests fully timed, see what you get wrong, and learn from those mistakes. Remember . . . it's good to make mistakes on these exams because if you learn from those mistakes now, you won't make them again in May!

The Graphics Used in This Book

To emphasize particular skills and strategies, we use several icons throughout this book. An icon in the margin will alert you that you should pay particular attention to the accompanying text. We use three icons:

1. This icon points out a very important concept or fact that you should not pass over.

2. This icon calls your attention to a problem-solving strategy that you may want to try.

3. This icon indicates a tip that you might find useful.

Boldfaced words indicate terms that are included in the Glossary at the end of the book. Boldface is also used to indicate the answer to a sample problem discussed in the test. Throughout the book you will find marginal notes, boxes, and starred areas. Pay close attention to these areas because they can provide tips, hints, strategies, and further explanations to help you reach your full potential.

THE FOUR "BIG IDEAS" OF AP BIOLOGY

The College Board has identified four key concepts, called "big ideas," that form the basis for the AP Biology curriculum. All the topics covered in the AP Biology course and all the questions on the AP Biology exam are linked to at least one of these big ideas.

The outline that follows provides a basic summary of the four "big ideas" of AP Biology. Each of the four ideas is subdivided into topics, and under each topic are statements that identify essential knowledge. You won't need to memorize this—you won't be asked to list even the big ideas—but you will need to be sure you have a good understanding of them. They are the fundamental concepts on which the course and the exam are based.

As you work through the content review chapters of this book, you'll find icons in the margin with cryptic numbers and letters. Don't let these confuse you! Those icons are there to help you understand how the subject matter in the review chapters fits into the big ideas of the AP Biology course. Look at the example in the left margin.

In this example, the number 4 refers to the fourth big idea, "Interactions" (see outline). The letter A refers to topic A under big idea number 4, which is "Biological systems (from cells to ecosystems) contain parts that interact with each other." The number 1 refers to the first point of essential knowledge under topic A, "The properties of biological molecules are determined by their components (monomers and polymers)." As you make your way through the subject review in this book, it is very helpful to keep an eye on the big picture and keep in mind how all the content you review relates to these four big ideas.

> **BIG IDEA 4.A.1**
> *The subcomponents of a molecule determine its properties.*

Big Idea 1: Evolution

A. Evolution is change over time.
1. Evolution occurs because of natural selection.
2. Natural selection acts on phenotypic variations in populations.
3. Evolutionary change is random.
4. Evolution is supported by scientific evidence.

B. Descent from common ancestry links organisms.
1. Many features are widely conserved (found in all sorts of organisms).
2. Phylogenetic trees and cladograms are visual representations of these relationships.

C. Evolution continues in a changing environment.
1. Speciation increases diversity and extinction reduces diversity; both have occurred throughout history.
2. Speciation is often a result of reproductive isolation.
3. Organisms continue to evolve.

D. Natural processes explain the origin of life.
1. There are different hypotheses, each supported by evidence, about how life began.
2. Different hypotheses of how life began are supported by scientific evidence from many disciplines.

Big Idea 2: Cellular Processes: Energy, Communication, and Homeostasis

A. Free energy and matter are required for life processes.
1. All living things need a constant source of energy.
2. Autotrophs (through photosynthesis) and heterotrophs (through cellular respiration) capture and store free energy.
3. Organisms exchange matter with the environment.

B. Cells maintain an internal environment that is different from their surroundings.
1. The structure of cell membranes results in selective permeability.
2. Homeostasis is maintained by movements through cell membranes.
3. Internal membranes compartmentalize cells (organelles).

C. Organisms rely on feedback mechanisms to grow, reproduce, and survive.
1. Organisms use positive and negative feedback.
2. Feedback mechanisms allow organisms to respond to changes in their environments.

D. Growth and homeostasis are influenced by environmental changes.
1. Biotic and abiotic interactions affect everything from cells to ecosystems.
2. Both common ancestry and divergence (due to different environments) are reflected in homeostatic mechanisms.
3. Disruptions to homeostasis affect the health of the organism and the balance of the ecosystem.
4. Plants and animals have defense systems against infection (immune systems).

E. Temporal (time-based) coordination is involved in homeostasis, growth, and reproduction.
1. The correct timing of events is necessary for the proper development of an organism.
2. The timing and coordination of the events is regulated in different ways for different organisms.
3. Various mechanisms are used to time and coordinate behavior.

Big Idea 3: Genetics and Information Transfer

A. The continuity of life is made possible through heritable information.
1. DNA and RNA provide heritable information from one generation to the next.
2. The cell cycle, mitosis, and meiosis are processes that transmit heritable information.
3. The transmittal of genes from parent to child shows the chromosomal basis of inheritance.
4. The inheritance of traits is more complex than can be explained by simple Mendelian models.

B. Cellular and molecular mechanisms are involved in gene expression.
1. Differential gene expression is a result of gene regulation.
2. Signal transmissions influence gene expression.

C. Imperfect processing of genetic information creates genetic variation.
1. Changes in genotype create new phenotypes.
2. Many biological mechanisms can increase variation.
3. Viral infections can produce genetic variation in their host organisms.

D. Cells communicate with other cells through the transmission of chemical signals.
1. Cell communication is based on common features (which show a common ancestry).

2. Cells can communicate through direct contact or long-distance signaling.
3. Signal transduction pathways begin with the recognition of a signal by a cell and ends with the cellular response.
4. A blocked or defective signal transduction pathway can change the cellular response.

E. Information transmission results in changes in biological systems.
1. Organisms communicate and exchange information.
2. Animal nervous systems perform a key role in the reception, transmittal, and processing of information to produce a response.

Big Idea 4: Interactions

A. Biological systems (from cells to ecosystems) contain parts that interact with each other.
1. The properties of biological molecules are determined by their components (monomers and polymers).
2. Organelles are essential to cell processes.
3. Interactions between external stimuli and gene expression produce specialized cells, tissues, and organs.
4. Complex properties in organisms are the result of interactions among their constituent parts.
5. Populations of organisms interact in communities.
6. The movement of matter and flow of energy are a result of interaction between communities and their environment.

B. Biological systems are characterized by competition and cooperation.
1. The structure and function of a molecule are influenced by its interaction with other molecules.
2. Cooperative interaction within organisms promotes efficiency.
3. Interactions within and between populations affect species distribution and abundance.
4. Ecosystem distribution changes over time (human impact accelerates change).

C. Diversity within biological systems influences interactions with the environment.
1. Variation in molecular units (e.g., different chlorophyll molecules, antibody proteins, or alleles of a gene) provides cells with a broad range of functions.
2. Gene expression is influenced by environment.
3. Variation influences population dynamics.
4. Diversity affects ecosystem stability.

STEP 1

Set Up Your Study Program

CHAPTER 1

What You Need to Know About the AP Biology Exam

IN THIS CHAPTER

Summary: Learn what topics are tested, how the test is scored, and basic test-taking information.

Key Ideas
- ✪ Some colleges will award credit for a score of 4 or 5.
- ✪ Multiple-choice and grid-in questions account for 50 percent of your final score.
- ✪ Points are no longer deducted for incorrect answers to multiple-choice questions. You should try to eliminate incorrect answer choices and then guess; there is no penalty for guessing.
- ✪ Free-response questions account for 50 percent of your final score.
- ✪ Your composite score on the two test sections is converted into a score on the 1-to-5 scale.

Background of the Advanced Placement Program

The Advanced Placement program was begun by the College Board in 1955 to construct standard achievement exams that would allow highly motivated high school students the opportunity to be awarded advanced placement as first-year students in colleges and universities in the United States. Today, more than a million students from every state in the nation and from foreign countries take the annual AP exams in May.

The AP programs are designed for high school students who wish to take college-level courses. In our case, the AP Biology course and exam are designed to involve high school students in college-level biology studies.

Who Writes the AP Biology Exam

After extensive surfing of the College Board website, here is what we have uncovered. The AP Biology exam is created by a group of college and high school biology instructors known as the AP Development Committee. The committee's job is to ensure that the annual AP Biology exam reflects what is being taught and studied in college-level biology classes at high schools.

This committee writes a large number of multiple-choice questions, which are pretested and evaluated for clarity, appropriateness, and range of possible answers. The committee also generates a pool of essay questions, pretests them, and chooses those questions that best represent the full range of the scoring scale, which will allow the AP readers to evaluate the essays equitably.

It is important to remember that the AP Biology exam is thoroughly evaluated after it is administered each year. This way, the College Board can use the results to make course suggestions and to plan future tests.

The AP Grades and Who Receives Them

Once you have taken the exam and it has been scored, your test will be graded with one of five numbers by the College Board:

- A 5 indicates that you are extremely well qualified.
- A 4 indicates that you are well qualified.
- A 3 indicates that you are adequately qualified.
- A 2 indicates that you are possibly qualified.
- A 1 indicates that you are not qualified to receive college credit.

A grade of 5, 4, 3, 2, or 1 will usually be reported by early July.

Reasons for Taking the AP Biology Exam

Why put yourself through a year of intensive study, pressure, stress, and preparation? Only you can answer that question. Following are some of the reasons that students have indicated to us for taking the AP exam:

- For personal satisfaction.
- To compare themselves with other students across the nation.
- Because colleges look favorably on the applications of students who elect to enroll in AP courses.
- To receive college credit or advanced standing at their colleges or universities.
- Because they love the subject.
- So that their families will be really proud of them.

There are plenty of other reasons, but no matter what they might be, the primary reason for enrolling in the AP Biology course and taking the exam in May is to feel good about yourself and the challenges you have met.

Questions Frequently Asked About the AP Biology Exam

Here are some common questions students have about the AP Biology exam and some answers to those questions.

If I Don't Take an AP Biology Course, Can I Still Take the AP Biology Exam?

Yes. Although the AP Biology exam is designed for students who have had a year's course in AP Biology, some high schools do not offer this type of course. Many students in these high schools have also done well on the exam, although they had not taken the course. However, if your high school does offer an AP Biology course, by all means take advantage of it and the structured background it will provide you.

How Is the Advanced Placement Biology Exam Organized?

The exam has two parts and is scheduled to last three hours. The first section is a set of 63 multiple-choice questions and six grid-in (calculation-based) questions. You will have 90 minutes to complete this part of the test.

After you complete the multiple-choice section, you will hand in your test booklet and scan sheet, and you will be given a brief break. The length of this break depends on the particular administrator. You will not be able to return to the multiple-choice questions when you return to the examination room.

The second section of the exam is a 90-minute essay-writing segment consisting of two long free-response questions and six short free-response questions. This section will be split into a 10-minute reading period followed by an 80-minute writing period. All of the questions will test your understanding of the four big ideas in biology and how science investigators actually work.

Must I Check the Box at the End of the Essay Booklet That Allows AP Staff to Use My Essays as Samples for Research?

No. This is simply a way for the College Board to make certain they have your permission if they decide to use one or more of your essays as a model. The readers of your essays pay no attention to whether or not that box is checked. Checking the box will not affect your grade.

How Is the Multiple-Choice Section Scored?

The scan sheet with your answers is run through a computer, which counts the number of correct answers. The AP Biology questions usually have four choices. A question left blank receives a zero. The very complicated formula for this calculation looks something like this (where N = the number of answers):

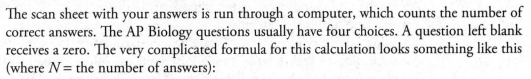

$$N_{right} = \text{raw score}$$

OK, that is not complicated at all.

How Are My Free-Response Answers Scored?

Each of your essays is read by a different, trained AP reader called a *faculty consultant*. The AP/College Board members have developed a highly successful training program for their readers, providing many opportunities for checks and double checks of essays to ensure a fair and equitable reading of each essay.

The scoring guides are carefully developed by a chief faculty consultant, a question leader, table leaders, and content experts. All faculty consultants are then trained to read and score just *one* essay question on the exam. They actually become experts in that one essay question. No one knows the identity of any writer. The identification numbers and names are covered, and the exam booklets are randomly distributed to the readers in packets of 25 randomly chosen essays. Table leaders and the question leader review samples of each reader's scores to ensure that quality standards are constant.

Each essay is scored on a scale from 1 to 10. Once your essay is graded on this scale, the next set of calculations is completed.

How Is My Composite Score Calculated?

This is where fuzzy math comes into play. The folks at the College Board are constantly adjusting the exact formula used to determine your composite score. Here is an example of a conversion they have used in the past, and this will be the conversion you will use to score any tests you do in this book.

The composite score for the AP Biology exam is 138. The free-response section represents 50 percent of this score, which equals 69 points. The multiple-choice section makes up 50 percent of the composite score, which equals another 69 points.

Take your multiple-choice results and plug them into the following formula (keep in mind that this formula was designed for a previous AP Biology exam and could be subject to some minor tweaking by the AP Board):

Number multiple-choice correct + number of grid-in correct = ———————

Take your essay results and plug them into this formula:

Total free-response points × 1.57 = ———————

Your total composite score for the exam is determined by adding the score from the multiple-choice section to the score from the essay section and rounding that sum to the nearest whole number.

How Is My Composite Score Turned into the Grade That Is Reported to My College?

Keep in mind that the total composite scores needed to earn a 5, 4, 3, 2, or 1 change each year. These cutoffs are determined by a committee of AP, College Board, and Educational Testing Service (ETS) directors, experts, and statisticians. The same exam that is given to the AP Biology high school students is given to college students. The various college professors report how the college students fared on the exam. This provides information for the chief faculty consultant on where to draw the lines for a 5, 4, 3, 2, or 1 score. A score of 5 on this AP exam is set to represent the average score received by the college students who scored an A on the exam. A score of a 3 or a 4 is the equivalent of a college grade B, and so on.

Over the years there has been an observable trend indicating the number of points required to achieve a specific grade. Data released from a particular AP Biology exam show that the approximate range for the five different scores are as follows (this changes from year to year—just use this as an approximate guideline):

- Mid 80s to 138 points = 5
- Mid 60s to lower 80s points = 4
- Upper 40s to lower 60s points = 3
- Mid 20s to upper 40s points = 2
- 0 to mid 20s points = 1

What Should I Bring to the Exam?

Here are some suggestions:

- A simple calculator
- Several pencils and an eraser
- Several black pens (black ink is easier on the eyes)
- A watch
- Something to drink—water is best
- A quiet snack, such as Lifesavers
- Your brain
- Tissues

What Should I *Avoid* Bringing to the Exam?

You should not bring:

- A jackhammer
- Loud stereo
- Pop rocks
- Your parents

Is There Anything Else I Should Be Aware Of?

You should:

- Allow plenty of time to get to the test site.
- Wear comfortable clothing.
- Eat a light breakfast or lunch.
- Remind yourself that you are well prepared and that the test is an enjoyable challenge and a chance to share your knowledge. Be proud of yourself! You worked hard all year. Once test day comes, there is nothing further you can do. It is out of your hands, and your only job is to answer as many questions correctly as you possibly can.

What Should I Do the Night Before the Exam?

Although we do not vigorously support last-minute cramming, there may be some value to some last-minute review. Spending the night before the exam relaxing with family or friends is helpful for many students. Watch a movie, play a game, gab on the phone, and then find a quiet spot to study. While you're unwinding, flip through your notebook and review sheets. As you are approaching the exam, you might want to put together a list of topics that have troubled you and review them briefly the night before the exam. If you are unable to fall asleep, flip through our chapter on taxonomy and classification (Chapter 13). Within moments, you're bound to be ready to drift off. Pleasant dreams.

CHAPTER 2

How to Plan Your Time

IN THIS CHAPTER

Summary: What to study for the AP Biology exam, depending on how much time you have available, plus three schedules to help you plan your course of study.

Key Ideas

- ✪ Focus your attention and spend time on those topics that are most likely to increase your score.
- ✪ Study the topics that you are *afraid* will appear, and relax about those that you know best.
- ✪ Do not study so widely that you forget to learn the important details of some of the more heavily detailed topics that appear on the AP Biology exam.

Three Approaches to Preparing for the AP Biology Exam

Overview of the Three Plans

No one knows your study habits, likes, and dislikes better than you do. So you are the only one who can decide which approach you want or need to adopt to prepare for the Advanced Placement Biology Exam. Look at the brief profiles below. These may help you determine a prep mode.

You're a Full-Year Prep Student (Plan A) if

1. You are the kind of person who likes to plan for everything very far in advance.

2. You arrive at the airport two hours before your flight because "you never know when these planes might leave early."

3. You like detailed planning and everything in its place.

4. You feel that you must be thoroughly prepared.

5. You hate surprises.

You're a One-Semester Prep Student (Plan B) if

1. You get to the airport one hour before your flight is scheduled to leave.

2. You are willing to plan ahead to feel comfortable in stressful situations, but are okay with skipping some details.

3. You feel more comfortable when you know what to expect, but a surprise or two is cool.

4. You're always on time for appointments.

You're a Six-Week Prep Student (Plan C) if

1. You get to the airport just as your plane is announcing its final boarding.

2. You work best under pressure and tight deadlines.

3. You feel very confident with the skills and background you've learned in your AP Biology class.

4. You decided late in the year to take the exam.

5. You like surprises.

6. You feel okay if you arrive 10–15 minutes late for an appointment.

General Outline of Three Different Study Plans

MONTH	PLAN A: FULL SCHOOL YEAR	PLAN B: ONE SEMESTER	PLAN C: SIX WEEKS
September–October	Introduction to material	—	—
November	Chapters 5–7	—	—
December	Chapters 8–9	—	—
January	Chapters 10–11	Chapters 5–7	—
February	Chapters 12–13	Chapters 8–10	—
March	Chapters 14–16	Chapters 11–14	—
April	Chapters 17–19; Practice Exam 1	Chapters 15–19; Practice Exam 1	Skim Chapters 5–14; all Rapid Review sections; Practice Exam 1
May	Review everything; Practice Exam 2	Review everything; Practice Exam 2	Skim Chapters 15–19; Practice Exam 2

Calendar for Each Plan

Plan A: You Have a Full School Year to Prepare

Although its primary purpose is to prepare you for the AP Biology exam you will take in May, this book can enrich your study of biology, your analytical skills, and your scientific essay-writing skills.

SEPTEMBER–OCTOBER (Check off the activities as you complete them.)

— Determine the study mode (A, B, or C) that applies to you.

— Carefully read Steps 1 and 2 of this book.

— Pay close attention to your walk-through of the Diagnostic Exam.

— Get on the web and take a look at the AP website(s).

— Skim the comprehensive review section (Step 4). (Reviewing the topics covered in this section will be part of your yearlong preparation.)

— Buy a few color highlighters.

— Flip through the entire book. Break the book in. Write in it. Toss it around a little bit . . . highlight it.

— Get a clear picture of what your own school's AP Biology curriculum is.

— Begin to use the book as a resource to supplement your classroom learning.

NOVEMBER (the first 10 weeks have elapsed)

— Read and study Chapter 5, Chemistry.

— Read and study Chapter 6, Cells.

— Read and study Chapter 7, Respiration.

DECEMBER

— Read and study Chapter 8, Photosynthesis.

— Read and study Chapter 9, Cell Division.

— Review Chapters 5–7.

JANUARY (20 weeks have elapsed)

— Read and study Chapter 10, Heredity.

— Read and study Chapter 11, Molecular Genetics.

— Review Chapters 5–9.

FEBRUARY

— Read and study Chapter 12, Evolution.

— Read and study Chapter 13, Taxonomy and Classification.

— Review Chapters 5–11.

MARCH (30 weeks have now elapsed)

— Read and study Chapter 14, Plants.

— Read and study Chapter 15, Human Physiology.

— Read and study Chapter 16, Human Reproduction.

— Review Chapters 5–13.

APRIL

— Take Practice Exam 1 in the first week of April.

— Evaluate your strengths and weaknesses.

— Study appropriate chapters to correct your weaknesses.

— Read and study Chapter 17, Behavioral Ecology and Ethology.

— Read and study Chapter 18, Ecology in Further Detail.

— Read and study Chapter 19, Laboratory Review.

— Review Chapters 5–16.

MAY (first 2 weeks) (THIS IS IT!)

— Review Chapters 5–19—all the material!

— Take Practice Exam 2.

— Score yourself.

— Get a good night's sleep before the exam. Fall asleep knowing that you are well prepared.

GOOD LUCK ON THE TEST!

Plan B: You Have One Semester to Prepare

Working under the assumption that you've completed one semester of biology studies,
the following calendar will use those skills you've been practicing to prepare
you for the May exam.

JANUARY
— Carefully read Steps 1 and 2 of this book.
— Take the Diagnostic Exam.
— Pay close attention to your walk-through of the Diagnostic Exam.
— Read and study Chapter 5, Chemistry.
— Read and study Chapter 6, Cells.
— Read and study Chapter 7, Respiration.

FEBRUARY
— Read and study Chapter 8, Photosynthesis.
— Read and study Chapter 9, Cell Division.
— Read and study Chapter 10, Heredity.
— Review Chapters 5–7.

MARCH (10 weeks to go)
— Read and study Chapter 11, Molecular Genetics.
— Read and study Chapter 12, Evolution.
— Review Chapters 8–10.
— Read and study Chapter 13, Taxonomy and Classification.
— Read and study Chapter 14, Plants.

APRIL
— Take Practice Exam 1 in the first week of April.
— Evaluate your strengths and weaknesses.

— Study appropriate chapters to correct your weaknesses.
— Read and study Chapter 15, Human Physiology.
— Review Chapters 5–9.
— Read and study Chapter 16, Human Reproduction.
— Read and study Chapter 17, Behavioral Ecology and Ethology.
— Review Chapters 10–14.
— Read and study Chapter 18, Ecology in Further Detail.
— Read and study Chapter 19, Laboratory Review.

MAY (first 2 weeks) (THIS IS IT!)
— Review Chapters 5–19, all the material!
— Take Practice Exam 2.
— Score yourself.
— Get a good night's sleep before the exam. Fall asleep knowing that you are well prepared.

GOOD LUCK ON THE TEST!

Plan C: You Have Six Weeks to Prepare

At this point, we assume that you have been building your biology knowledge
base for more than six months. You will, therefore, use this book primarily
as a specific guide to the AP Biology exam.
Given the time constraints, now is not the time to try to expand your
AP Biology curriculum. Rather, you should focus on and refine
what you already know.

APRIL 1–15
— Skim Steps 1 and 2 of this book.
— Skim Chapters 5–9.
— Carefully go over the Rapid Review sections of Chapters 5–9.
— Complete Practice Exam 1.
— Score yourself and analyze your errors.
— Skim and highlight the Glossary at the end of the book.

APRIL 16–MAY 1
— Skim Chapters 10–14.
— Carefully go over the Rapid Review sections of Chapters 10–14.

— Carefully go over the Rapid Review sections for Chapters 5–9.
— Continue to skim and highlight the Glossary.

MAY (first 2 weeks) (THIS IS IT!)
— Skim Chapters 15–19.
— Carefully go over the Rapid Review sections of Chapters 15–19.
— Complete Practice Exam 2.
— Score yourself and analyze your errors.
— Get a good night's sleep. Fall asleep knowing that you are well prepared.

GOOD LUCK ON THE TEST!

STEP **2**

Determine Your Test Readiness

CHAPTER **3**   Take a Diagnostic Exam

CHAPTER 3

Take a Diagnostic Exam

IN THIS CHAPTER

Summary: In the following pages you will find a diagnostic exam. It is intended to give you an idea of your level of preparation in biology. After you have completed the test, check your answers against the given answers.

Key Ideas

❂ Practice the kind of multiple-choice questions you might be asked on the real exam.

❂ Answer questions that approximate the coverage of themes on the real exam.

❂ Check your work against the given answers.

❂ Determine your areas of strength and weakness.

❂ Highlight the concepts to which you must give special attention.

Answer Sheet for Diagnostic Exam in AP Biology

PART A: MULTIPLE-CHOICE QUESTIONS

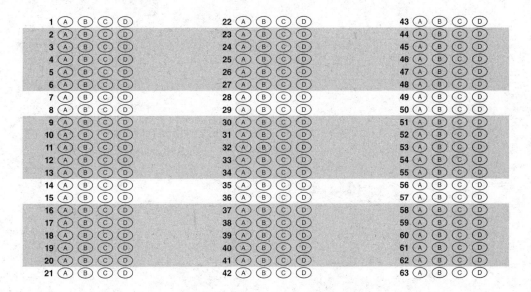

PART B: GRID-IN QUESTIONS

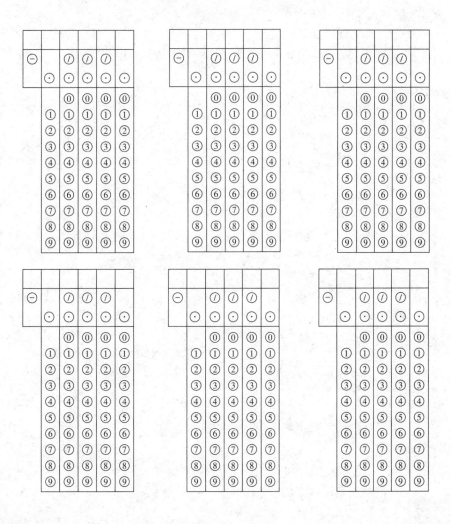

DIAGNOSTIC EXAM: AP BIOLOGY: SECTION I

PART A: MULTIPLE-CHOICE QUESTIONS

Time—1 hour and 30 minutes (for Parts A and B)
For the multiple-choice questions that follow, select the best answer and fill in the appropriate letter on the answer sheet.

1. A pH of 10 is how many times more basic than a pH of 7?

   A. 10
   B. 100
   C. 1,000
   D. 10,000

2. Destruction of microfilaments would most adversely affect which of the following?

   A. Cell division
   B. Cilia
   C. Flagella
   D. Muscular contraction

3. Imagine that for a particular species of moth, females are primed to respond to two types of male mating calls. Males who produce an in-between version will not succeed at obtaining a mate and will therefore have low reproductive success. This is an example of

   A. directional selection.
   B. stabilizing selection.
   C. artificial selection.
   D. disruptive selection.

4. Crossover occurs during

   A. prophase of mitosis.
   B. prophase I of meiosis.
   C. prophase II of meiosis.
   D. prophase I and II of meiosis.

5. Which of the following is a specialized feature of plants that live in hot and dry regions?

   A. Stomata that open and close
   B. Transpiration
   C. Photophosphorylation
   D. C_4 photosynthesis

6. A virus that carries the reverse transcriptase enzyme is

   A. a retrovirus.
   B. a prion.
   C. a viroid.
   D. a DNA virus.

7. Ants live on acacia trees and are able to feast on the sugar produced by the trees. The tree is protected by the ants' attack on any foreign insects that may harm the tree. This is an example of

   A. parasitism.
   B. commensualism.
   C. mutualism.
   D. symbiosis.

8. In which of the following structures would one most likely find smooth muscle?

   A. Biceps muscle
   B. Digestive tract
   C. Quadriceps muscle
   D. Gluteus maximus muscle

9. Halophiles would be classified into which major kingdom?

   A. Monera
   B. Protista
   C. Plantae
   D. Fungi

10. A reaction that breaks down compounds by the addition of water is known as

    A. a hydrolysis reaction.
    B. a dehydration reaction.
    C. an endergonic reaction.
    D. an exergonic reaction.

11. In humans, the developing embryo tends to attach to this structure.

    A. Fallopian tube
    B. Oviduct
    C. Endometrium
    D. Cervix

12. Plants that produce a single spore type that gives rise to bisexual gametophytes are called

    A. heterosporous.
    B. gymnosperms.
    C. homosporous.
    D. angiosperms.

13. In humans, spermatogenesis, the process of male gamete formation, occurs in the

    A. seminiferous tubules.
    B. epididymis.
    C. vas deferens.
    D. seminal vesicles.

14. Which of the following is an example of aneuploidy?

    A. Cri-du-chat syndrome
    B. Chronic myelogenous leukemia
    C. Turner syndrome
    D. Achondroplasia

15. Among the following choices, which one would most readily move through a selectively permeable membrane?

    A. Small, uncharged polar molecule
    B. Large, uncharged polar molecule
    C. Glucose
    D. Sodium ion

16. Which of the following is not a lipid?

    A. Steroid
    B. Fat
    C. Phospholipid
    D. Glycogen

17. Which of the following hormones is *not* released by the anterior pituitary gland?

    A. Follicle-stimulating hormone (FSH)
    B. Antidiuretic hormone (ADH)
    C. Growth hormone (GH or STH)
    D. Adrenocorticotropic hormone (ACTH)

18. Which of the following is the *least* specific taxonomic classification category?

    A. Division
    B. Order
    C. Family
    D. Genus

19. These cells control the opening and closing of a plant's stomata:

    A. Guard cells
    B. Collenchyma cells
    C. Parenchyma cells
    D. Mesophyll cells

20. Imagine that 9 percent of a population of anteaters have a short snout (recessive), while 91 percent have a long snout (dominant). If this population is in Hardy–Weinberg equilibrium, what is the expected frequency (in percent) of the heterozygous condition?

    A. 30.0
    B. 34.0
    C. 38.0
    D. 42.0

21. The situation in which a gene at one locus alters the phenotypic expression of a gene at another locus is known as

    A. incomplete dominance.
    B. codominance.
    C. pleiotropy.
    D. epistasis.

22. The oxygen produced during the light reactions of photosynthesis comes directly from

    A. H_2O.
    B. H_2O_2.
    C. $C_2H_3O_2$.
    D. CO_2.

23. An organism that alternates between a haploid and a diploid multicellular stage during its life cycle is most probably a

    A. shark.
    B. human.
    C. pine tree.
    D. amoeba.

24. The presence of which of the following organelles or structures would most convincingly indicate that a cell is a eukaryote and not a prokaryote?

    A. Plasma membrane
    B. Cell wall
    C. Lysosome
    D. Ribosome

25. Traits that are similar between organisms that arose from a common ancestor are known as

    A. convergent characters.
    B. homologous characters.
    C. vestigial characters.
    D. divergent characters.

26. The process by which a huge amount of DNA is created from a small amount of DNA in a very short amount of time is known as

A. cloning.
B. transformation.
C. polymerase chain reaction.
D. gel electrophoresis.

27. A compound contains a COOH group. What functional group is that?

A. Carbonyl group
B. Carboxyl group
C. Hydroxyl group
D. Phosphate group

28. Which of the following forms of cell transport requires the input of energy?

A. Diffusion
B. Osmosis
C. Facilitated diffusion
D. Active transport

29. Homologous chromosomes are chromosomes that

A. are found only in identical twins.
B. are formed during mitosis.
C. split apart during meiosis II.
D. resemble one another in shape, size, and function.

30. Which of the following is an incorrect statement about DNA replication?

A. It occurs in the nucleus.
B. It occurs in a semiconservative fashion.
C. Helicase is the enzyme that adds the nucleotides to the growing strand.
D. DNA polymerase can build only in a 5′-to-3′ direction.

31. Warning coloration adopted by animals that possess a chemical defense mechanism is known as

A. cryptic coloration.
B. deceptive markings.
C. aposematic coloration.
D. batesian mimicry.

32. In a large pond that consists of long-finned fish and short-finned fish, a tornado wreaks havoc on the pond, killing 50 percent of the fish population. By chance, most of the fish killed were short-finned varieties, and in the subsequent generation there were fewer fish with short fins. This is an example of

A. gene flow.
B. bottleneck.
C. balanced polymorphism.
D. allopatric speciation.

33. Which of the following structures would not have developed from the mesoderm?

A. Muscle
B. Heart
C. Kidneys
D. Liver

34. Which of the following is not a characteristic of bryophytes?

A. They were the first land plants.
B. They contain a waxy cuticle to protect against water loss.
C. They package their gametes into gametangia.
D. The dominant generation is the sporophyte.

35. The cyclic pathway of photosynthesis occurs because

A. the Calvin cycle uses more ATP than NADPH.
B. it can occur in regions lacking light.
C. it is a more efficient way to produce oxygen.
D. it is a more efficient way to produce the NADPH needed for the Calvin cycle.

36. Which of the following conditions is an X-linked condition?

A. Hemophilia
B. Tay-Sachs disease
C. Cystic fibrosis
D. Sickle cell anemia

37. The uptake of foreign DNA from the surrounding environment is known as

    A. generalized transduction.
    B. specialized transduction.
    C. conjugation.
    D. transformation.

38. Most of the digestion of food occurs in the

    A. esophagus.
    B. stomach.
    C. small intestine.
    D. large intestine.

39. You have just come back from visiting the redwood forests in California and were amazed at how *wide* those trees were. What process is responsible for the increase in width of these trees?

    A. Growth of guard cells
    B. Growth of collenchyma cells
    C. Growth of apical meristem cells
    D. Growth of lateral meristem cells

40. The trophoblast formed during the early stages of human embryology eventually develops into the

    A. placenta.
    B. embryo.
    C. hypoblast.
    D. morula.

41. What biome is known for having the greatest diversity of species?

    A. Taiga
    B. Temperate grasslands
    C. Tropical forest
    D. Savanna

42. In hypercholesterolemia, a genetic condition found in humans, individuals who are HH have normal cholesterol levels, those who are hh have horrifically high cholesterol levels, and those who are Hh have cholesterol levels that are somewhere in between. This is an example of

    A. dominance.
    B. incomplete dominance.
    C. codominance.
    E. epistasis.

43. The light-dependent reactions of photosynthesis occur in the

    A. nucleus.
    B. cytoplasm.
    C. thylakoid membrane.
    D. stroma.

44. Which of the following is a characteristic of an R-selected strategist?

    A. Low reproductive rate
    B. Extensive postnatal care
    C. Relatively constant population size
    D. J-shaped growth curve

45. Which of the following statements about mitosis is correct?

    A. Mitosis makes up 30 percent of the cell cycle.
    B. The order of mitosis is prophase, anaphase, metaphase, telophase.
    C. Single-cell eukaryotes undergo mitosis as part of asexual reproduction.
    D. Cell plates are formed in animal cells during mitosis.

46. A vine that wraps around the trunk of a tree is displaying the concept known as

    A. photoperiodism.
    B. thigmotropism.
    C. gravitropism.
    D. phototropism.

47. This hormone is known for assisting in the closing of the stomata, and inhibition of cell growth.

    A. Abscisic acid
    B. Cytokinin
    C. Ethylene
    D. Gibberellin

48. Antigen invader → B-cell meets antigen → B-cell differentiates into plasma cells and memory cells → plasma cells produce antibodies → antibodies eliminate antigen. The preceding sequence of events is a description of

    A. cell-mediated immunity.
    B. humoral immunity.
    C. nonspecific immunity.
    D. cytotoxic T-cell maturation.

For questions 49–51, please use the following answers:

A. Abscisic acid
B. Cytokinins
C. Ethylene
D. Gibberellins

49. This hormone is known to promote cell division in plant roots and shoots.

50. This hormone is known to regulate stem elongation, germination, flowering, and other developmental processes.

51. This hormone is produced in the roots of a plant in response to decreased soil water potential and other situations in which the plant may be under stress.

For questions 52–55, please use the following answers:

A. Glycolysis
B. Oxidative phosphorylation
C. Chemiosmosis
D. Fermentation

52. This reaction occurs in the mitochondria and involves the formation of ATP from NADH and $FADH_2$.

53. The coupling of the movement of electrons down the electron transport chain with the formation of ATP using the driving force provided by the proton gradient.

54. This reaction occurs in the cytoplasm and has as its products 2 ATP, 2 NADH, and 2-pyruvate.

55. This reaction is performed by cells in an effort to regenerate the NAD^+ required for glycolysis to continue.

For questions 56–59, please use the following answer choices:

A. Associative learning
B. Insight learning
C. Imprinting
D. Altruistic behavior

56. The ability to reason through a problem the first time through with no prior experience.

57. Action in which an organism helps another, even if it comes at its own expense.

58. Process by which an animal substitutes one stimulus for another to get the same response.

59. Innate behavior learned during a critical period early in life.

For questions 60–63, please use the information from the following laboratory experiment:

You are working as a summer intern at the local university laboratory, and a lab technician comes into your room, throws a few graphs and tables at you, and mutters, "Interpret this data for me . . . I need to go play golf. I'll be back this afternoon for your report." Analyze the data this technician so kindly gave to you, and use it to answer questions 60–63. The reaction rates reported in the tables are relative to the original rate of the reaction in the absence of the enzymes. The three enzymes used are all being added to the same reactants to determine which should be used in the future.

Room Temperature (25°C), pH 7

ENZYME	REACTION RATE
1	1.24
2	1.51
3	1.33

Varying Temperature, Constant (pH 7)

ENZYME	0°C	5°C	10°C	15°C	20°C	25°C	30°C	35°C	40°C
1	1.00	1.02	1.04	1.19	1.20	1.24	1.29	1.27	1.22
2	1.01	1.12	1.35	1.39	1.65	1.51	1.40	1.12	1.01
3	1.06	1.21	1.55	1.44	1.35	1.33	1.15	1.10	1.06

Varying pH, Constant Temperature = 25°C

ENZYME	4	5	6	7	8	9	10
1	1.54	1.51	1.33	1.24	1.20	1.08	1.05
2	1.75	1.71	1.62	1.51	1.32	1.10	1.01
3	1.52	1.45	1.40	1.33	1.20	1.09	1.04

60. If you had also been given a graph that plotted the moles of product produced versus time, what would have been the best way to calculate the rate for the reaction?

A. Calculate the average of the slope of the curve for the first and last minute of reaction.
B. Calculate the slope of the curve for the portion of the curve that is constant.
C. Calculate the slope of the curve for the portion where the slope begins to flatten out.
D. Add up the total number of moles produced during each time interval and divide by the total number of time intervals measured.

61. Over the interval measured, at what temperature does enzyme 2 appear to have its optimal efficiency?

A. 10°C
B. 15°C
C. 20°C
D. 25°C

62. Which of the following statements about enzyme 3 is incorrect?

A. At a pH of 6 and a temperature of 25°C, it is more efficient than enzyme 2 but less efficient than enzyme 1.
B. It functions more efficiently in the acidic pH range than the basic pH range.
C. At 30°C and a pH of 7, it is less efficient than both enzymes 1 and 2.
D. Over the interval given, its optimal temperature at a pH of 7 is 10°C.

63. Which of the following statements can be made from review of these data?

A. Enzyme 1 functions most efficiently in a basic environment and at a lower temperature.
B. Enzyme 1 functions more efficiently than enzyme 2 at 10°C and a pH of 7.
C. The pH does not affect the efficiency of enzyme 3.
D. All three enzymes function more efficiently in an acidic environment than a basic environment.

PART B: GRID-IN QUESTIONS

Calculate the correct answer and enter it on the top line of the grid-in area with each number/symbol in a separate column. Then fill in the correct circle below each number/symbol you entered (only one filled-in circle per column).

1. Sickle cell anemia is a mutation in hemoglobin that affects the shape of red blood cells during periods of low oxygenation. Sickle cell anemia displays recessive inheritance. An expecting mother and father visit a geneticist for counseling. Both parents are carriers of the sickle cell trait. Calculate the likelihood that their child is a carrier.

2. Dwarfism is an autosomal dominant disease. A couple has sought out genetic counseling as they prepare to start a family. They want to have two biological children. One parent has been diagnosed with dwarfism (Dd); the second parent is healthy (dd). Calculate the probability that *both* children will have dwarfism.

3. A population of mice is in Hardy-Weinberg equilibrium. The recessive phenotype is found in 1 out of every 2500 mice. Calculate the percentage of the heterozygous phenotype.

4. A population of fruit flies is in Hardy-Weinberg equilibrium. The allele for black eyes (B) is dominant to the red allele (b). The recessive phenotype is seen in 36% of the population. Calculate the frequency of the dominant allele.

5. In a certain breed of dog, the allele for red-colored tongue (T) is dominant to the allele for purple tongue (t). A researcher collected data on 48 dogs bred from a cross between a red-tongued dog and a purple-tongued dog. Of the 48 off-spring, 36 had red tongues and 12 had purple tongues. Calculate the chi-squared value for the null hypothesis, assuming that the red-tongued dog was heterozygous for the tongue-color gene.

Chi-Square Significance Table

DEGREE OF FREEDOM (n)	5% PROBABILITY VALUE (P)
1	3.84
2	5.99
3	7.81
4	9.49

6. **pH and Hydrogen ion calculation**

The pH of liquid A is 3. Calculate the hydrogen ion concentration.

$$pH = -\log[H+]$$

AP Biology Diagnostic Exam: Section II

FREE–RESPONSE QUESTIONS

Time—1 hour and 30 minutes

(The first 10 minutes is a reading period. Do not begin writing until the 10–minute period has passed.)
Questions 1 and 2 are long free-response questions that should require about 20 minutes each.
Questions 3–8 are short-response questions that should require about 6 minutes each.
Outline form is not acceptable. Answers should be in essay form.

1. Inheritance, Pedigrees, and Calculations
Huntington's disease is a genetic illness that leads to degeneration of the central nervous system. Symptoms typically do not present until between 30 and 40 years of age.

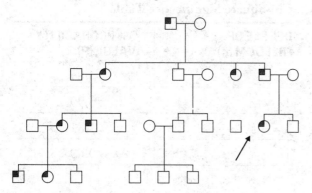

A. A pedigree is shown at the left. The grey marks indicate that the individual has Huntington's. What type of inheritance pattern is suspected? Explain your reasoning.

B. The couple indicated by the arrow are planning to have a child. Assume the husband is healthy. Create a Punnett square to demonstrate the risk of inheritance for their offspring.

C. A randomized trial is run to examine whether "Drug Lauder" delayed the emergence of symptoms among individuals who carried the genetic trait for inheritance. Age at first emergence of symptoms is shown under the patient.

DRUG STATUS	PATIENT 1	PATIENT 2	PATIENT 3	PATIENT 4	PATIENT 5	PATIENT 6
Control	32	29	41			
Lauder, 50mg				47	46	42

On the axis provided below, create an appropriately labeled bar graph of the average age of symptom onset of the two populations.

Average age of symptom onset by years of age

_____

_____

_____

_____

_____

2. Digestion, Cellular Respiration, and Evolution
Adequate nutritional intake, along with absorption of nutrients, is necessary for bodily functioning.

A. Cellular respiration, comprised of glycolysis, the Krebs cycle, and the electron transport chain, utilizes the glucose from food in order to synthesize ATP. The majority of ATP is created in the electron transport chain. Briefly explain how the electron transport chain creates ATP.

A researcher is interested in the small and large intestines. He has recognized a new syndrome that significantly reduces nutrient absorption.

B. Briefly differentiate the functions of the small and large intestines.

C. If a patient's small intestine is completely removed, hypothesize the impact that this could have on his weight.

DRUG STATUS	PATIENT 1	P2	P3	P4	P5	P6	P7	P8	MEAN NUTRIENT ABSORPTION LEVEL
Control	50	43	76	34					50.75
New Drug					100	78	76	87	85.25

The researcher runs a pilot study to determine whether a newly developed drug is effective in treating patients with significantly reduced nutrient absorption. Nutrient absorption levels for each patient and the means are shown above.

D. Interpret the preliminary results shown.

E. Suggest one limitation of the above study, or propose one variable that should be controlled if the study is repeated.

3. Meiosis is a form of cell division in all sexually reproducing organisms during which the nucleus divides during the production of spores or gametes.

A. What are two aspects of meiosis that contribute to genetic diversity?

B. The mating of two individuals who are closely related is termed *consanguineous*. How would consanguinity decrease the genetic diversity of offspring, and why might this decreased diversity be harmful to the offspring?

4. Homeostasis, or the maintenance of balance within the body, is necessary to normal functioning.

A. Explain how insulin and glucagon provide an example of homeostasis.

B. How do the concepts of blood glucose level, insulin, and negative feedback relate?

5. Scientists have recently discovered that the raccoon population in rural New York communicates using chemical signals, via pheromones. The scientists consequently developed four pheromones in an attempt to replicate the natural pheromone. They measured the response time of five raccoons to the developed chemical compounds as an indication of chemical similarity between the developed and natural pheromones. The response times are provided, along with the average response time and standard error of the mean for each developed pheromone.

A. Animals communicate through several mechanisms. Briefly describe TWO methods of communication.

B. Based on the mean response times and the standard error of the mean, explain which pheromone most closely aligned to the raccoon's naturally produced pheromones.

C. A follow up study on raccoon pheromone response time is planned to determine whether response to pheromones is an innate or learned response. Response times will be compared between baby raccoons, young raccoons, and adult raccoons. Hypothesize what the data will show if the response to pheromones is an innate response.

PHEROMONE	RACCOON 1	RACCOON 2	RACCOON 3	RACCOON 4	RACCOON 5	MEAN RESPONSE TIME	STANDARD ERROR OF THE MEAN (SEM)
I	15	13	16	7	8	11.8	1.8
II	10	8	7	11	15	10.2	1.4
III	3	6	5	9	4	5.4	1.0

6. A young girl presents to the emergency department with symptoms suggesting the effects of a bacterial pathogen. Her parents report that they did not vaccinate their daughter against this particular pathogen.
  A. Briefly describe TWO components of the innate/primary immune response.
  B. If the child had received a vaccination for this particular pathogen, how would the adaptive/secondary immune response differ? Reference the graph shown below.

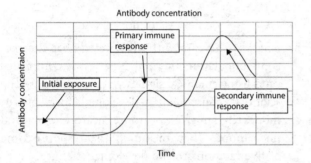

7. A population of foxes and rabbits in Environment A exhibits the expected predator–prey population curve seen below.
  A. Explain the relationship represented by the curve.

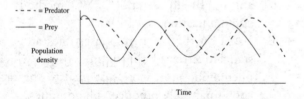

  B. A bacterial pathogen completely wipes out the population of foxes. Researchers closely watch the rabbit population and notice a new population curve. Explain this new curve. In your explanation, be sure to explain the fluctuations seen at the new carrying capacity.

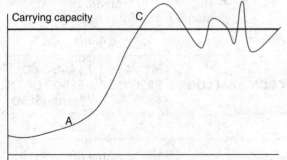

8. A population of wild cats lives in a temperate environment with four seasons. The cats have three main fur colors—white, light brown, and black.

A scientist decides to transplant a random sample of these cats to an environment covered in snow throughout the full year. He notices that as time passes, the cats with one fur color tend to live longer and evade predators better than those with the other two fur colors.
  A. Identify the cats with the fur color you expect to live longer in this new environment and explain your reasoning.
  B. The scientist is also interested in better understanding the evolutionary relationships between the wild cats and several other wild species. He collected data on various traits. Construct a visual representation of the evolutionary relationships from the data provided.

Species	TRAIT A	TRAIT B	TRAIT C
Animal 1	+	−	−
Animal 2	+	−	+
Animal 3	+	+	+
Animal 4	−	−	−

〉 Answers and Explanations

PART A: MULTIPLE-CHOICE QUESTIONS

1. **C**—This question deals with the concept of pH: acids and bases. The pH scale is a logarithmic scale that measures how acidic or basic a solution is. A pH of 4 is 10 times more acidic than a pH of 5. A pH of 6 is 10^2 or 100 times more basic than a pH of 4, and so on. Therefore, a pH of 10 is 10^3 or 1000 times more basic than a pH of 7.

2. **D**—This question deals with the cytoskeleton of cells. Cell division, cilia, and flagella would be compromised if the *microtubules* were damaged. Microfilaments, made from actin, are important to muscular contraction. Chitin is a polysaccharide found in fungi.

3. **D**—This is a prime example of disruptive selection. Take a look at the material from Chapter 12 on the various types of selection. The illustrations there are worth reviewing.

4. **B**—You have to know this fact. We don't want them to get you on this one if they ask it. ☺

5. **D**—C_4 photosynthesis is an adaptive photosynthetic process that attempts to counter the problems that hot and dry weather causes for plants. Be sure that you read about and understand the various forms of photosynthesis for the exam.

6. **A**—Retroviruses are RNA viruses that carry with them the reverse transcriptase enzyme. When they take over a host cell, they first use the enzyme to convert themselves into DNA. They next incorporate into the DNA of the host, and begin the process of viral replication. The HIV virus of AIDS is a well-known retrovirus.

7. **C**—Mutualism is the interaction in which both parties involved benefit.

8. **B**—Smooth muscle is found in the digestive tract, bladder, and arteries, to name only a few. Answer choices A and D are skeletal muscles.

9. **A**—Halophiles are a member of the archaebacteria subgroup of the monerans.

10. **A**—This question deals with five types of reactions you should be familiar with for the AP Biology exam. A hydrolysis reaction is one in which water is added, causing the formation of a compound.

11. **C**—Fertilization tends to occur in the oviduct, also known as the *fallopian tube*. The ovum is produced in the ovary, and the cervix is the passageway from the uterus to the vagina.

12. **C**—Homosporous plants, such as ferns, give rise to bisexual gametophytes.

13. **A**—You should learn the general processes of spermatogenesis and oogenesis in humans for the AP Biology exam.

14. **C**—Turner syndrome (XO) is an example of aneuploidy—conditions in which individuals have an abnormal number of chromosomes. These conditions can be monosomies, as is the case with Turner, or they can be trisomies, as is the case with Down, Klinefelter, and other syndromes.

15. **A**—The selectively permeable membrane is a lipid bilayer composed of phospholipids, proteins, and other macromolecules. Small, uncharged polar molecules and lipids are able to pass through these membranes without difficulty.

16. **D**—Glycogen is a carbohydrate. The three major types of lipids you should know are fats, phospholipids, and steroids. Cholesterol is a type of steroid.

17. **B**—This hormone, which is involved in controlling the function of the kidney, is released from the posterior pituitary.

18. **A**—The stupid phrase we use to remember this classification hierarchy is "Karaoke players can order free grape soda"—kingdom, phylum, class, order, family, genus, and species. This question is sneaky because it requires you to know that a division is the plant kingdom's version of the phylum. The kingdom is the least specific subdivision, and the species the most specific. Therefore, A is the correct answer.

19. **A**—Guard cells are the cells responsible for controlling the opening and closing of the stomata of a plant.

20. **D**—If 9 percent of the population is recessive (ss), then $q^2 = 0.09$. Taking the square root of 0.09 gives us $q = 0.30$. Knowing as we do that $p + q = 1$, $p + 0.30 = 1$, and $p = 0.70$. The frequency of the heterozygous condition $= 2pq = 2(0.30)(0.70) = 42\%$.

21. **D**—Epistasis exists when a gene at one locus affects a gene at another locus.

22. **A**—The inputs to the light reactions include light and water. During these reactions, photolysis occurs, which is the splitting of H_2O into hydrogen ions and oxygen atoms. These oxygen atoms from the water pair together immediately to form the oxygen we breathe.

23. **C**—This life cycle is the one known as "alternation of generations." It is the plant life cycle. Pine trees are the only ones among the choices that would show such a cycle.

24. **C**—Prokaryotes are known for their simplicity. They do not contain a nucleus, nor do they contain membrane-bound organelles. They do have a few structures to remember: cell wall, plasma membrane, ribosomes, and a nucleoid. Lysosomes are found in eukaryotes, not prokaryotes.

25. **B**—Traits are said to be homologous if they are similar because their host organisms arose from a common ancestor. For example, the bone structure in bird wings is homologous in all bird species.

26. **C**—Polymerase chain reaction is the high-speed cloning machine of molecular genetics. It occurs at a much faster rate than does cloning.

27. **B**—Functional groups are a pain in the neck. But you need to be able to recognize them on the exam. Most often, the test asks students to identify functional groups by structure.

28. **D**—Active transport requires energy. The major types of cell transport you need to know for the exam are diffusion, osmosis, facilitated diffusion, endocytosis, exocytosis, and active transport.

29. **D**—Homologous chromosomes resemble one another in shape, size, and function. They pair up during meiosis and separate from each other during meiosis I.

30. **C**—DNA polymerase is the superstar enzyme of the replication process, which occurs during the S phase of the cell cycle in the nucleus of a cell. The process does occur in semiconservative fashion. You should learn the basic concepts behind replication as they are explained in Chapter 11.

31. **C**—Learn the defense mechanisms well from predator–prey relationships in Chapter 18. They will be represented on the exam.

32. **B**—A bottleneck is a specific example of genetic drift: the sudden change in allele frequencies due to random events.

33. **D**—You should learn the list of structures derived from endoderm, mesoderm, and ectoderm. (This could be an easy multiple-choice question for you if you do.)

34. **D**—The dominant generation for bryophytes is the gametophyte (n) generation. They are the only plants for which this is true.

35. **A**—The Calvin cycle uses a disproportionate amount of ATP relative to NADPH. The cyclic light reactions exist to make up for this disparity. The cyclic reactions do not produce NADPH, nor do they produce oxygen.

36. **A**—Tay-Sachs disease, cystic fibrosis, and sickle cell anemia are all autosomal recessive conditions. It will serve you well to learn the most common autosomal recessive conditions, X-linked conditions, and autosomal dominant conditions.

37. **D**—It will serve you well for this exam to be reasonably familiar with biotechnology laboratory techniques. Lab procedures show up often on free-response questions and the later multiple-choice sections of the exam.

38. **C**—The small intestine hosts the most digestion of the digestive tract.

39. **D**—This is known as *cambium*.

40. **A**—The inner cell mass gives rise to the embryo, which eventually gives rise to the epiblast and hypoblast. The morula is an early stage of development.

41. **C**—Biomes are annoying and tough to memorize. Learn as much as you can about them without taking up too much time. . . . More often than not there will be two to three multiple-choice questions about them. But you want to make sure you learn enough to work your way through a free-response question if you were to be so unfortunate as to have one on your test.

42. **B**—Incomplete dominance is the situation in which the heterozygous genotype produces an "intermediate" phenotype rather than the dominant phenotype; neither allele dominates the other.

43. **C**—The light-dependent reactions occur in the thylakoid membrane. The dark reactions, known as the *Calvin cycle*, occur in the stroma.

44. **D**—A J-shaped growth curve is characteristic of exponentially growing populations. That is a characteristic of R-selected strategists.

45. **C**—Mitosis makes up 10 percent of the cell cycle; the correct order of the stages is prophase, metaphase, anaphase, telophase; mitosis is not performed by prokaryotic cells; and cell plates are formed in plant cells.

46. **B**—Thigmotropism, phototropism, and gravitropism are the major tropisms you need to know for plants. Thigmotropism, the growth response of a plant to touch, is the least understood of the bunch.

47. **A**—There are five plant hormones you should know for the exam. Auxin seems to come up the most, but it would serve you well to know the basic functions of all five of them.

48. **B**—Humoral immunity is another name for antibody-mediated immunity. Cell-mediated immunity involves T-cells and the direct cellular destruction of invaders such as viruses.

49. **B**

50. **D**

51. **A**

52. **B**—Each NADH is able to produce up to 3 ATP. Each $FADH_2$ can produce up to 2 ATP.

53. **C**—You have to know the concept of chemiosmosis for the AP exam. Make sure you study it well in Chapter 7.

54. **A**—Glycolysis is the conversion of glucose into pyruvate that occurs in the cytoplasm and is the first step of both aerobic and anaerobic respiration.

55. **D**—Fermentation is anaerobic respiration, and it is the process that begins with glycolysis and ends with the regeneration of NAD^+.

56. **B**—Chapter 17 is fairly short and concise. We left it to the bare bones for you to learn. We would suggest you learn this chapter well because it could be worth a good 5–7 points for you on the exam if you are lucky. ☺

57. **D**

58. **A**

59. **C**

60. **B**—The rate of reaction for an enzyme-aided reaction is best estimated by taking the slope of the constant portion of the moles–time plot.

61. **C**—They will test your ability to interpret data on this exam. You should make sure that you are able to look at a chart and interpret information given to you. This enzyme does indeed function most efficiently at 20°C. Above and below that temperature, the reaction rate is lower.

62. **A**—At a pH of 6 and a temperature of 25°C, enzyme 3 is actually LESS efficient than enzyme 2 and MORE efficient than enzyme 1.

63. **D**—This question requires you to know that a pH below 7 (pH < 7) is acidic and a pH above 7 (pH > 7) is basic. It is true that all three enzymes increase the rate of reaction more when in acidic environments than basic environments.

PART B: GRID–IN QUESTIONS

1. 0.50

Let H^A = healthy and H^C = sickle cell trait

	H^A	H^C
H^A	$H^A\ H^A$	$H^A\ H^C$
H^C	$H^A\ H^C$	$H^C\ H^C$

The child will have a 25% chance of being healthy, a 50% chance of being a carrier, and a 25% chance of having sickle cell anemia.

2. 0.25

First, determine the risk for one child of inheriting dwarfism. Since this is an autosomal dominant disease, the child needs to inherit only one dominant allele (Dd) to express the disease.

	D	d
d	Dd	dd
d	Dd	dd

The Punnett square demonstrates that each child has a 50% (or ½) chance of inheriting the dominant allele. The couple wants to know the probability that their two children BOTH will have dwarfism. To calculate the probability that both children inherit the dominant allele, multiply the probability of each individual event.

$$\frac{1}{2} * \frac{1}{2} = \frac{1}{4}$$

Thus the probability of both children being diagnosed with dwarfism is 25%, or ¼.

3. .04

There are two main Hardy–Weinberg equations: $p + q = 1$ AND $p^2 + 2pq + q^2 = 1$

p and q represent the frequency of the dominant and recessive alleles respectively. p^2 and q^2 represent the frequency of the homozygous dominant and homozygous recessive phenotypes respectively. Lastly, $2pq$ represents the frequency of the heterozygous phenotype.

Hardy–Weinberg problems often start by providing information about the recessive phenotype. In this case, we know that the recessive phenotype is in 1 out of every 2500 mice; we can turn this into a frequency by creating a fraction: $1/2500 = 0.0004$. This frequency is the homozygous recessive phenotype; thus we can equate $q^2 = 0.0004$.

The question asks for the frequency of the heterozygous phenotype (represented by $2pq$ above). Using our finding $q^2 = 0.0004$, we can determine the value of $2pq$.

$$q^2 = 0.0004$$

so $q = 0.02$

To calculate p, we can use the equation: $p + q = 1$
Plugging in $q = 0.02 \rightarrow p + 0.02 = 1 \rightarrow p = 0.98$

Now to calculate $2pq$ (frequency of the heterozygous phenotype), plug in the known values: $2(0.98)(0.02) = 0.04$. In other words, 4% of the population is heterozygous.

The frequency of the dominant allele is thus 40%.

4. 0.40

Much of the logic from the previous problem will apply here. To reiterate, there are two main Hardy–Weinberg equations: $p + q = 1$ AND $p^2 + 2pq + q^2 = 1$. p and q represent the frequency of the dominant and recessive alleles respectively. p^2 and q^2 represent the frequency of the homozygous dominant and homozygous recessive phenotypes respectively. Lastly, $2pq$ represents the frequency of the heterozygous phenotype.

If 36% of the population is the recessive phenotype, $q^2 = .36$; thus, $q = 0.6$

We want the value of p, the dominant allele. We can use: $p + q = 1$

With $q = 0.6 \rightarrow p + 0.6 = 1 \rightarrow p = 0.4$

5. 12

We are given the observed values in the question prompt. Now you need to calculate the expected phenotypes utilizing a Punnett square. The question stem has provided that the cross is between a heterozygous red–tongued dog (Tt) and a purple–tongued dog (tt).

	T	t
t	Tt	tt
t	Tt	tt

The Punnett square suggests that this cross would lead to 50% red tongue, 50% white tongue. Out of 48 dogs, 24 would have a red tongue, and 24 would have a purple tongue.

Phenotype	# Observed (o)	# Expected (e)	o − e	$\dfrac{(o - e)^2}{e}$
Red tongue	36	24	12	$\dfrac{(12)^2}{24} = 6$
Purple tongue	12	24	−12	$\dfrac{(-12)^2}{24} = 6$
TOTAL	48	48		SUM = 12

For two options, the degree of freedom is 1. Since the chi–squared value of 12 is larger than the 5% probability value of 3.84, the data does not follow predicted values and is statistically significant. Discard the null hypothesis.

				1	2
⊖		/	/	/	
	⊙	⊙	⊙	⊙	⊙
		⓪	⓪	⓪	⓪
	①	①	①	●	①
	②	②	②	②	●
	③	③	③	③	③
	④	④	④	④	④
	⑤	⑤	⑤	⑤	⑤
	⑥	⑥	⑥	⑥	⑥
	⑦	⑦	⑦	⑦	⑦
	⑧	⑧	⑧	⑧	⑧
	⑨	⑨	⑨	⑨	⑨

6. .001

$$pH = -\log [H+]$$
$$3 = -\log [H+]$$
$$10-3 = [H+]$$
$$0.001 = [H+]$$

		.	0	0	1
⊖		/	/	/	
	●	⊙	⊙	⊙	⊙
	⓪	●	●	⓪	
	①	①	①	①	●
	②	②	②	②	②
	③	③	③	③	③
	④	④	④	④	④
	⑤	⑤	⑤	⑤	⑤
	⑥	⑥	⑥	⑥	⑥
	⑦	⑦	⑦	⑦	⑦
	⑧	⑧	⑧	⑧	⑧
	⑨	⑨	⑨	⑨	⑨

❭ Free–Response Grading Outline

1. A. Autosomal dominance inheritance is suspected. Looking at the pedigree, each generation is affected by the illness, which supports dominant as opposed to recessive inheritance. Recessive inheritance would instead show frequent skipping of generations in the pedigree. Autosomal as opposed to sex–linked inheritance is suspected, since males and females are equally affected.

B. To create the Punnett square, we need to determine whether the individual in row 5 is homozygous or heterozygous for the trait. Since one of her parents is unaffected, we can assume she is heterozygous. We are also told that her husband is healthy and, thus, homozygous recessive.

	H	h
h	Hh	hh
h	Hh	hh

The Punnett square above suggests that each offspring maintains a 50% chance of inheriting the disease.

C. First calculate the average age of symptom onset for each of the two conditions:

$$\text{Control} = \frac{32 + 29 + 41}{3} = 34 \text{ yrs old}$$

$$\text{Drug} = \frac{47 + 46 + 42}{3} = 45 \text{ yrs old}$$

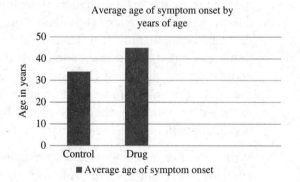

Average age of symptom onset by years of age

2. A. The electron transport chain creates a proton gradient which sends protons (H^+) out of the mitochondrial matrix using the energy–carrying NADH and $FADH_2$. This proton is then allowed to flow back down through the ATP synthase and create the ATP necessary for bodily function.

B. The small intestine, comprised of the duodenum, jejunum, and ileum, is where the bulk of digestion and absorption occurs. The large intestine, comprised of the cecum, colon, and rectum, has the function of reabsorbing water from digested material and working to eliminate waste.

C. Since the small intestine is the primary site of nutrient absorption, if the small intestine is completely removed, the patient would consequently experience inadequate nutrition. Subsequently, we would be worried that the patient's weight would decrease.

D. The preliminary results suggest that the drug has higher levels of absorption than the control.

E. A limitation of the study is the small sample size. A few variables that should be controlled include age/sex/ethnicity of patients, time that the study drug is administered, dosage of the study drug, etc.

3. A. Two processes in meiosis that contribute to genetic diversity:
  1. *Crossing over* between homologous chromosomes during prophase I. Complementary DNA strands are exchanged between the chromosomes.
  2. *Random assortment* of homologous chromosomes at the metaphase plate. Homologous chromosomes are divided in half to become haploid cells. It is by chance on which side each pair of chromosomes from the mother and father align.

B. Impact of consanguinity:
1. Genetic diversity would decrease since the mother and father maintain a more similar genetic makeup than a nonconsanguineous pairing.
2. Decreased diversity in this case could prove harmful in the case of recessive traits. For example, a trait could be rare, but since the mother and father are related, they could be more likely to both be carriers. If both parents are carriers, the probability that the child inherits one or two recessive alleles is increased.

4. A. Insulin and glucagon are hormones produced in the pancreas that work to control the level of glucose in the blood. Insulin is released from the pancreatic beta cells when glucose levels are higher in the blood—insulin stimulates cellular uptake of glucose for utilization or glycogen production. On the other hand, glucagon is released from the pancreatic alpha cells when glucose levels are lower in the blood. Glucagon stimulates the release of glycogen from the liver to raise glucose levels in the blood. The opposing effects of insulin and glucagon maintain the homeostasis of blood glucose.

B. Negative feedback occurs when a hormone inhibits further release of a hormone. Concerning the question, when blood glucose levels are high, insulin is released by the pancreas to stimulate cellular uptake of the glucose. The pancreas recognizes the consequent decrease in blood glucose and responds by decreasing or stopping the release of further insulin.

5. A. Communication types include:
• Visual: the use of visual cues to relay meaning. Examples include peacock feather coloring, bared teeth, rolling over on back.
• Tactile: the use of touch to relay meaning. For example, monkeys will groom each other as a sign of affection.
• Auditory: the use of various sounds to relay meaning. For example, the frog chirp to attract a mate.

• Chemical: the use of chemical signals via pheromones. Pheromones play a significant role in animal mating.

B. Pheromone III demonstrated the fastest mean of the three pheromones tested and thus would be more likely to resemble the natural raccoon pheromones.

C. An innate response is a behavior that does not need to be taught. Thus if the response to pheromones is innate, the response time of the young raccoons to the pheromones will be similar to the response time of the older raccoons.

6. A. Components of the innate immune response:
• Physical barriers like the skin, GI tract, respiratory tract, nose hairs
• Broad defenses like mucous secretions, stomach acid, saliva, tears
• Aspects of the general immune response like the complement system, phagocytosis of pathogens via macrophages and neutrophils, inflammation to call in additional immune support, NK cells

B. Secondary exposure
• Response time would be faster
• Indicated by the presence of memory cells
• Production of antibodies is greater in quantity, and they demonstrate higher affinity.

7. A. The curve demonstrates that the population concentrations are correlated. The foxes rely on the rabbits for food. As the foxes consume more rabbits, the fox population will grow due to higher supply levels. At some point, the rabbit population will have a higher death rate than birth rate, and thus the population will diminish. Consequently, the fox population will too decrease. Then with decreased predation by the foxes, the rabbit population will be able to rejuvenate.

B. With the fox population and predation removed, the rabbits will be able to sustain exponential population growth. At some point, the rabbit population will reach a new carrying capacity, the maximum population size that the environment can support, and thus will level off.

8. A. This is a type of directional selection, where one extreme of the phenotype is "safer" or more adaptive in the environment. In this situation, the environment has snow year around. Thus the cats that have white fur are able to blend into their environment better than the cats with black or brown fur and evade predation more successfully.

B. To create a cladogram, it is often easier to determine outliers for each trait. Beginning with Trait A, Animals 1–3 all have the trait, but Animal 4 does not.

Animal 4 also does not have Traits B and C. Next with Trait C, Animals 2 and 3 share the trait, but Animal 1 does not. Lastly with Trait B, only Animal 3 has the trait.

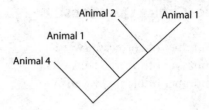

Scoring and Interpretation
AP BIOLOGY DIAGNOSTIC EXAM

Multiple–Choice Questions: **Grid–In Questions:**

number of correct answers: _____	number of correct answers: _____
number of incorrect answers: _____	number of incorrect answers: _____
number of blank answers: _____	number of blank answers: _____

_____ + _____ = _____
multiple–choice   grid–in   Section I
number correct   number correct   raw score

Free–Response Questions:

1. ____ / 10   3. ____ / 4   6. ____ / 4
2. ____ / 10   4. ____ / 4   7. ____ / 4
                 5. ____ / 4   8. ____ / 4

Add up the total points accumulated in the eight questions and multiply the sum by 1.57 to obtain the free–response raw score: _____ $\times 1.57 =$ _____
                         free–response points   Section II raw score

CALCULATE YOUR SCORE

Now combine the raw scores from the multiple–choice and free–response sections to obtain your new raw score for the entire practice exam. Use the ranges listed below to determine your grade for this exam. Don't worry about how we arrived at the following ranges, and remember that they are rough estimates on questions that are not actual AP exam questions . . . do not read too much into them.

Raw Score	Approximate AP Score
83–138	5
63–82	4
48–62	3
26–47	2
0–25	1

If this test went amazingly well for you . . . rock and roll . . . but as we just said, your journey is just beginning, and that means you have time to supplement your knowledge even more before the big day! Use your time well.

If this test went poorly for you, don't worry; as has been said twice now, your journey is just beginning and you have plenty of time to learn what you need to know for this exam. Just use this as an exercise in focus that has shown you what you need to concentrate on between now and early May. Good luck!

STEP 3

Develop Strategies for Success

CHAPTER **4** How to Approach Each Question Type

CHAPTER 4

How to Approach Each Question Type

IN THIS CHAPTER

Summary: Become familiar with the types of questions on the exam: multiple-choice and free-response. Pace yourself and know when to skip a question that you can come back to later.

Key Ideas

- ✪ On multiple-choice questions, you no longer lose any points for wrong answers. So you should bubble in an answer for *every* question.
- ✪ On multiple-choice questions, don't "out-think" the test. Use common sense because that will usually get you to the right answer.
- ✪ Free-response answers must be in essay form. Outline form is not acceptable.
- ✪ Free-response questions tend to be multi-part questions—be SURE to answer each part of the question or you will not be able to get the maximum possible number of points for that question.
- ✪ Make a quick outline before you begin writing your answer.
- ✪ The free-response questions are graded using a positive-scoring system, so wrong information is ignored.

Multiple-Choice Questions

You have approximately 90 seconds per question on the multiple-choice section of this exam. Remember that to be on pace for a great score on this exam, you need to correctly answer approximately 42 multiple-choice questions or more. Here are a few rules of thumb:

1. *Don't out-think the test.* It is indeed possible to be too smart for these tests. Frequently during these standardized tests we have found ourselves overanalyzing every single problem. If you encounter a question such as, "During what phase of meiosis does crossover (also referred to as *crossing over*) occur?" and you happen to know the answer immediately, this does not mean that the question is too easy. First, give yourself credit for knowing a fact. They asked you something, you knew it, and *wham*, you fill in the bubble. Do not overanalyze the question and assume that your answer is too obvious for that question. Just because you get it doesn't mean that it was too easy.

2. *Don't leave questions blank.* The AP Biology exam used to take off one-fourth point for each wrong answer. This is no longer the case. You should bubble in an answer for each multiple-choice question.

3. *Be on the lookout for trick wording!* Always pay attention to words or phrases such as "least," "most," "not," "incorrectly," and "does not belong." Do not answer the wrong question. There are few things as annoying as getting a question wrong on this test simply because you didn't read the question carefully enough, especially if you know the right answer.

4. *Use your time carefully.* Some of these questions require a lot of careful reading before you can answer them. If you find yourself struggling on a question, try not to waste too much time on it. Circle it in the booklet and come back to it later if time permits. Remember—you are looking to answer approximately 42 multiple-choice questions correctly to be on pace for a great score—this test should be an exercise in window shopping.

It does not matter *which* questions you get correct. What is important is that you answer enough questions correctly. Find the subjects that you know the best, answer those questions, and save the others for review later on.

5. *Be careful about changing answers!* If you have answered a question already, come back to it later on, and get the urge to change it . . . make sure that you have a real *reason* to change it. Often an urge to change an answer is the work of exam "elves" in the room who want to trick you into picking a wrong answer. Change your answer only if you can justify your reasons for making the switch.

6. *Check your calculations!* The math required in the grid-in section isn't overly complicated. That said, it would be unfortunate to lose points because of a silly calculation error. Make sure to work carefully and check your math. Happily, any equations you need will be provided for you.

Free-Response Questions

The free-response section consists of eight broad questions. It is important that your answers to these questions display solid reasoning and analytical skills. The two long essays together carry approximately the same weight as the six short-response questions combined. Expect to often use data or information from your laboratory exercises as you answer some of the questions.

Answers for the free-response questions must be in essay form. Outline form is not acceptable. Labeled diagrams may be used to supplement discussion, but in no case will a diagram alone suffice. It is important that you read each question completely before you begin to write. Write all of your answers on the pages following the questions in the booklet.

Free-Response Tips

Some important tips to keep in mind as you write your essays:

- The free-response questions tend to be multipart questions. You can't be expected to know everything about every topic, and the test preparers sometimes throw you a bone by writing questions that ask you to answer *two* of three parts or *three* of four parts. This gives you an opportunity to focus in on the material that you are most comfortable with. It is very important that you read the question carefully to make sure you understand exactly what the examiners are asking you to do.

- You are given 80 minutes to complete eight free-response questions. The two long free-response questions should take 20 minutes each, and the six short questions should take about six minutes each. This may not seem like a lot of time, but if you write a bunch of practice essays before you take the exam and budget your time wisely during the exam, you will not have to struggle with your timing. Below are suggestions for budgeting your time:
  - Read the question carefully and make sure you know what it is asking you to do.
  - Construct an outline that will help you organize your answer. Don't write the world's most elaborate outline. You won't get points for having the prettiest outline in the country—so there is no reason to spend an excessive amount of time putting it together. Just develop enough of an outline so that you have a basic idea of how you will construct your essay. Your essay is not graded based on how well it is put together, but it certainly will not hurt your score to write a well-organized and grammatically correct response.
  - If the long essay is a two-part question, spend 10 minutes on each part. If it is a three-part question, spend 6–7 minutes on each part. Keep your eye on the clock and make sure you give yourself enough time to address each part of the question.

- Both of the long free-response questions on the AP Biology exam are worth the same number of points. But each question is not created equal. Some questions ask you to answer two sub-questions. Some questions ask you to answer three sub-questions, and some questions ask you to answer four sub-questions. The free-response questions are graded in a way that forces you to provide information for *each* section of the question. There are a maximum number of points that you can get for each subsection. For example, in a question that asks you to answer *three* sub-questions, most likely the grader's guidelines will say something along the lines of:

> Part A — worth a maximum of 3 points
> Part B — worth a maximum of 4 points
> Part C — worth a maximum of 3 points

This is a very important thing for you to know heading into the exam. This means that it is *far* more important for you to attempt to answer every *part* of the question than to try to stuff every little fact that you know about part A into that portion of the essay at the expense of part B. Based on the grading guideline above, no matter how well you write your answer for part A, you can receive at most 3 points for that section. At the risk of being repetitive, we'll say it again because it is so important: no matter how great your essay may be, the grader can only give you the maximum possible number of points for each subsection.

- The free-response section is graded using a "positive scoring" system. This means that wrong information in an essay is ignored. You do not lose points for saying things that are incorrect. (Unfortunately you do not *get* points for saying things that are incorrect either . . . if only!) The importance of this fact is basically that if you are unsure about something and think you may be right, give it a shot and include it in your essay. It's worth the risk.

STEP 4

Review the Knowledge You Need to Score High

CHAPTER 5

Chemistry

IN THIS CHAPTER

Summary: This chapter introduces the chemical principles that are related to the AP Biology topics covered throughout the course.

Key Ideas

✪ Organic compounds contain carbon; important examples include lipids, proteins, and carbohydrates.

✪ Enzymes are catalytic proteins that react in an induced-fit fashion with substrates to speed up a reaction.

✪ The five types of chemical reactions you should learn include hydrolysis reactions, dehydration synthesis reactions, endergonic reactions, exergonic reactions, and redox reactions.

Introduction

What is the name of the test you are studying for? The AP Biology exam. Then why in tarnation are we starting your review with a chapter titled *Chemistry*?!?!? Because it is important that you have an understanding of a few chemical principles before we dive into the deeper biological material. We will keep it short, don't worry. ☺

Elements, Compounds, Atoms, and Ions

By definition, **matter** is anything that has mass and takes up space; an **element** is defined as matter in its simplest form; an **atom** is the smallest form of an element that still displays its particular properties. (Terms boldfaced in text are listed in the Glossary at the end of the book.)

For example, sodium (Na) is an element mentioned often in this book, especially in Chapter 15, Human Physiology. The element sodium can exist as an atom of sodium, in which it is a neutral particle containing an equal number of protons and electrons. It can also exist as an ion, which is an atom that has a positive or negative charge. Ions such as sodium that take on a positive charge are called **cations,** and are composed of more protons than electrons. Ions with a negative charge are called **anions,** and are composed of more electrons than protons.

Elements can be combined to form **molecules,** for example, an oxygen molecule (O_2) or a hydrogen molecule (H_2). Molecules that are composed of more than one type of element are called **compounds,** for example H_2O. The two major types of compounds you need to be familiar with are **organic** and **inorganic** compounds. Organic compounds contain carbon and usually hydrogen; inorganic compounds do not. Some of you are probably skeptical, at this point, as to whether any of what we have said thus far matters for this exam. Bear with me because it does. You will deal with many important organic compounds later on in this book, including **carbohydrates, proteins, lipids,** and **nucleic acids.**

Before moving onto the next section, where we discuss these particular organic compounds in more detail, we would like to cover a topic that many find confusing and therefore ignore in preparing for this exam. This is the subject of **functional groups.** These poorly understood groups are responsible for the chemical properties of organic compounds. They should not intimidate you, nor should you spend a million hours trying to memorize them in full detail. You should remember one or two examples of each group and be able to identify the functional groups on sight, as you are often asked to do so on the AP exam.

The following is a list of the functional groups you should study for this exam:

John (11th grade): "My teacher wanted me to know these structures . . . she was right!"

> **BIG IDEA 4.A.1**
> *The subcomponents of a molecule determine its properties.*

1. *Amino group.* An amino group has the following formula:

$$R-N \begin{matrix} H \\ H \end{matrix}$$

The symbol R stands for "rest of the compound" to which this NH_2 group is attached. One example of a compound containing an amino group is an **amino acid.** Compounds containing amino groups are generally referred to as **amines.** Amino groups act as bases and can pick up protons from acids.

2. *Carbonyl group.* This group contains two structures:

$$\begin{matrix} R \\ | \\ C=O \\ | \\ R \end{matrix} \qquad R-C \begin{matrix} O \\ H \end{matrix}$$

ketone   **aldehyde**

If the C=O is at the end of a chain, it is an **aldehyde.** Otherwise, it is a **ketone.** (*Note:* in *aldehydes,* there is an H at the end; there is no H in the word *ketone.*) A carbonyl group makes a compound **hydrophilic** and **polar.** *Hydrophilic* means water-loving, reacting well with water. A *polar* molecule is one that has an unequal distribution of charge, which creates a positive side and a negative side to the molecule.

3. **Carboxyl group.** This group has the following formula:

$$\begin{matrix} R \\ C \\ OH \end{matrix} O$$

A *carboxyl group* is a carbonyl group that has a hydroxide in one of the R spots and a carbon chain in the other. This functional group shows up along with amino groups in amino acids. Carboxyl groups act as acids because they are able to donate protons to basic compounds. Compounds containing carboxyl groups are known as *carboxylic acids*.

4. *Hydroxyl group.* This group has the simplest formula of the bunch:

$$R - OH$$

A hydroxyl group is present in compounds known as **alcohols.** Like carbonyl groups, hydroxyl groups are polar and hydrophilic.

5. *Phosphate group.* This group has the following formula:

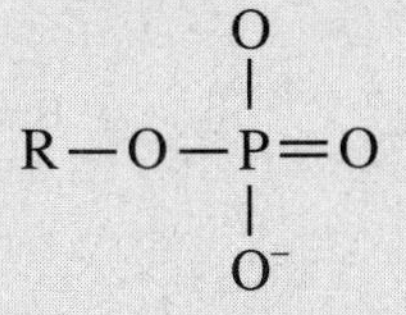

$$R - O - \overset{\displaystyle O}{\underset{\displaystyle O^-}{\overset{|}{\underset{|}{P}}}} = O$$

Phosphate groups are vital components of compounds that serve as cellular energy sources: ATP, ADP, and GTP. Like carboxyl groups, phosphate groups are acidic molecules.

6. *Sulfhydryl group.* This group also has a simple formula:

$$R - SH$$

This functional group does not show up much on the exam, but you should recognize it when it does. This group is present in the amino acids methionine and cysteine and assists in structure stabilization in many proteins.

Lipids, Carbohydrates, and Proteins

Lipids

BIG IDEA 4.C.1
These various molecules provide the cell with a wide range of functions.

Lipids are organic compounds used by cells as long-term energy stores or building blocks. Lipids are hydrophobic and insoluble in water because they contain a hydrocarbon tail of CH_2S that is nonpolar and repellant to water. The most important lipids are **fats, oils, steroids,** and **phospholipids.**

Fats, which are lipids made by combining **glycerol** and three **fatty acids** (Figure 5.1), are used as long-term energy stores in cells. They are not as easily metabolized as carbohydrates, yet

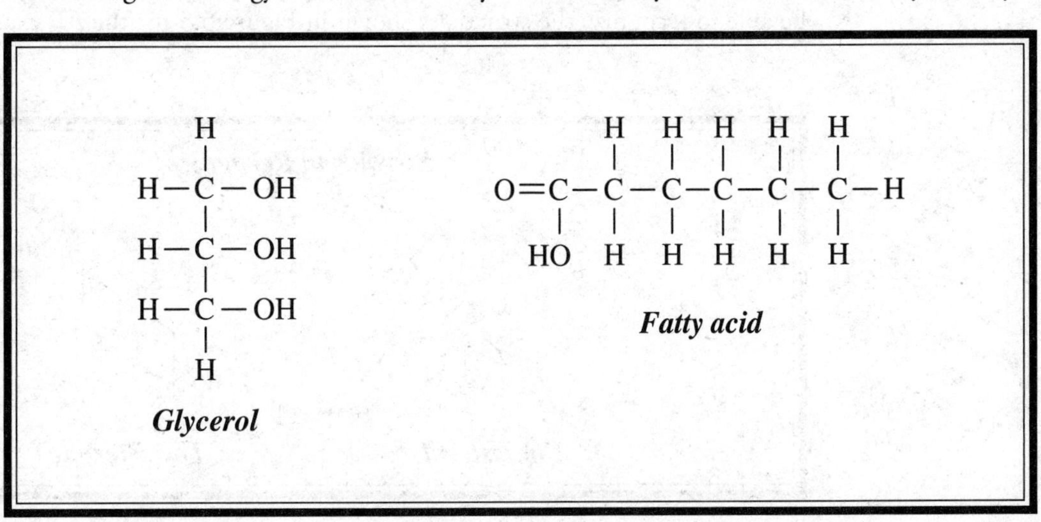

Figure 5.1   Structure of glycerol and fatty acids.

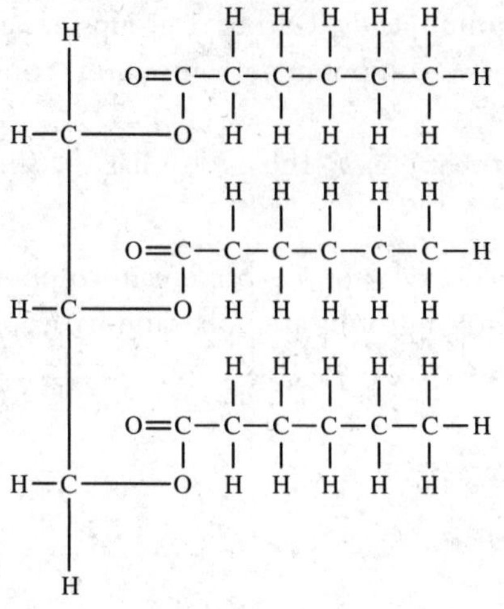

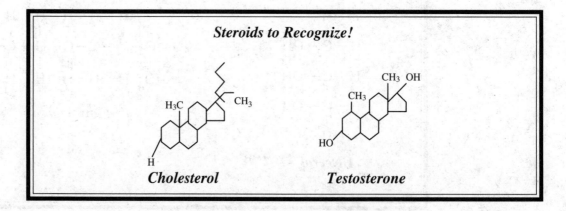

Figure 5.2   Fat structure (glycerol plus three fatty acids).

they are a more effective means of storage; for instance, one gram of fat provides two times the energy of one gram of carbohydrate. Fats can be **saturated** or **unsaturated**. Saturated fat molecules contain no double bonds. Unsaturated fats contain one (mono-) or more (poly-) double bonds, which means that they contain fewer hydrogen molecules per carbon than do saturated fats. Saturated fats are the bad guys and are associated with heart disease and atherosclerosis. Most of the fat found in animals is saturated, whereas plants tend to contain unsaturated fats. Fat is formed when three fatty-acid molecules connect to the OH groups of the glycerol molecule. These connecting bonds are formed by dehydration synthesis reaction (Figure 5.2).

Steroids are lipids composed of four carbon rings that look like chicken-wire fencing in pictorial representations. One example of a steroid is cholesterol, an important structural component of cell membranes that serves as a precursor molecule for another important class of steroids: the sex hormones (testosterone, progesterone, and estrogen). You should be able to recognize the structures shown in Figure 5.3 for the AP exam.

Steroids to Recognize!

Cholesterol   *Testosterone*

Figure 5.3   Steroid structures.

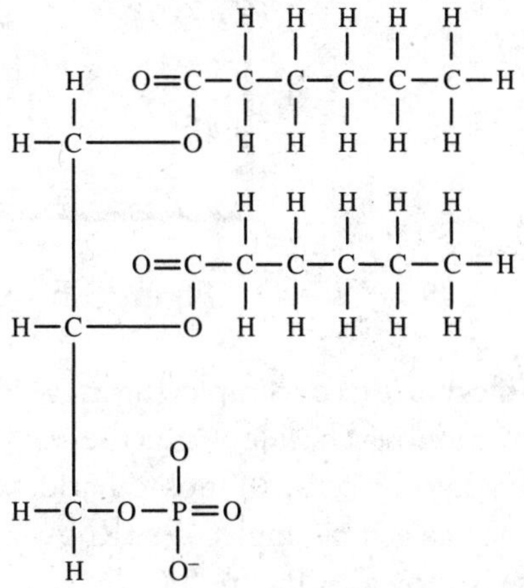

$$\begin{array}{ccccccc}
& & & \text{H} & \text{H} & \text{H} & \text{H} & \text{H} \\
& & & | & | & | & | & | \\
\text{H} & \text{O}{=}\text{C}{-}\text{C}{-}\text{C}{-}\text{C}{-}\text{C}{-}\text{H} \\
| & | & | & | & | & | \\
\text{H}{-}\text{C}{-\!-\!-\!-}\text{O} & \text{H} & \text{H} & \text{H} & \text{H}
\end{array}$$

Figure 5.4   Structure of phospholipid.

A **phospholipid** is a lipid formed by combining a glycerol molecule with two fatty acids and a phosphate group (Figure 5.4). Phospholipids are bilayered structures; they have both a hydrophobic tail (a hydrocarbon chain) and a hydrophilic head (the phosphate group) (Figure 5.5). They are the major component of cell membranes; the hydrophilic phosphate group forms the outside portion and the hydrophobic tail forms the interior of the wall.

Carbohydrates

Carbohydrates can be simple sugars or complex molecules containing multiple sugars. Carbohydrates are used by the cells of the body in energy-producing reactions and as structural materials. Carbohydrates have the elements C, H, and O. Hydrogen and oxygen are present in a 2:1 ratio. The three main types of carbohydrates you need to know are monosaccharides, disaccharides, and polysaccharides.

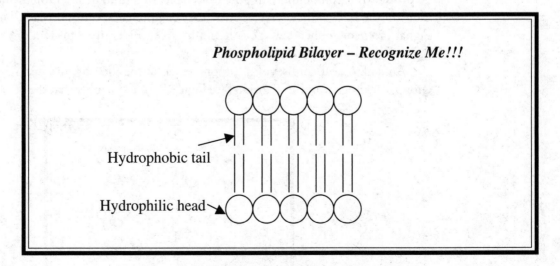

Figure 5.5   Bilayered structure of phospholipids.

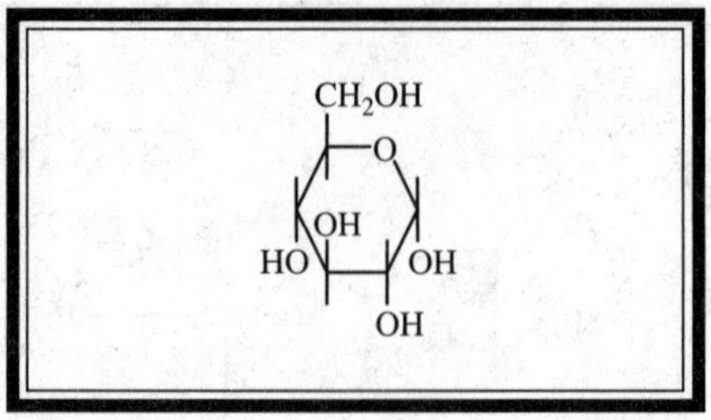

CH$_2$OH

Figure 5.6 Glucose structure.

A **monosaccharide,** or simple sugar, is the simplest form of a carbohydrate. The most important monosaccharide is glucose ($C_6H_{12}O_6$), which is used in cellular respiration to provide energy for cells. Monosaccharides with five carbons ($C_5H_{10}O_5$) are used in compounds such as genetic molecules (RNA) and high-energy molecules (ATP). The structure of glucose is shown in Figure 5.6.

A **disaccharide** is a sugar consisting of two monosaccharides bound together. Common disaccharides include sucrose, maltose, and lactose. Sucrose, a major energy carbohydrate in plants, is a combination of fructose and glucose; maltose, a carbohydrate used in the creation of beer, is a combination of two glucose molecules; and lactose, found in dairy products, is a combination of galactose and glucose.

A **polysaccharide** is a carbohydrate containing three or more monosaccharide molecules. Polysaccharides, usually composed of hundreds or thousands of monosaccharides, act as a storage form of energy and as structural material in and around cells. The most important carbohydrates for storing energy are **starch** and **glycogen**. Starch, made solely of glucose molecules linked together, is the storage form of choice for plants. Animals store much of their carbohydrate energy in the form of glycogen, which is most often found in liver and muscle cells. Glycogen is formed by linking many glucose molecules together.

Julie (11th grade): "Remembering these four came in handy on the test!"

Two important structural polysaccharides are **cellulose** and **chitin**. Cellulose, a compound composed of many glucose molecules, is used by plants in the formation of their cell walls. Chitin is an important part of the exoskeletons of arthropods such as insects, spiders, and shellfish (see Chapter 13, Taxonomy and Classification).

Proteins

A **protein** is a compound composed of chains of amino acids. Proteins have many functions in the body—they serve as structural components, transport aids, enzymes, and cell signals, to name only a few. You should be able to identify a protein or an amino acid by sight if asked to do so on the test.

An amino acid consists of a carbon center surrounded by an amino group, a carboxyl group, a hydrogen, and an R group (See Figure 5.7). Remember that the R stands for "rest" of

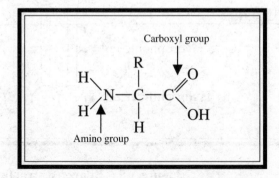

Figure 5.7 Structure of an amino acid.

$$H_2N-\underset{\underset{H}{|}}{\overset{\overset{R}{|}}{C}}-\underset{\underset{O}{\parallel}}{\overset{\overset{H}{|}}{C}}-\underset{\underset{H}{|}}{\overset{\overset{H}{|}}{N}}-\underset{\underset{H}{|}}{\overset{\overset{R}{|}}{C}}-\underset{\underset{O}{\parallel}}{\overset{\overset{H}{|}}{C}}-\underset{\underset{H}{|}}{\overset{\overset{H}{|}}{N}}-\underset{\underset{H}{|}}{\overset{\overset{R}{|}}{C}}-\underset{OH}{\overset{\overset{R}{|}}{C}}=O$$

Peptide bonds

Figure 5.8   Amino acid structure exhibiting peptide linkage.

the compound, which provides an amino acid's unique personal characteristics. For instance, acidic amino acids have acidic R groups, basic amino acids have basic R groups, and so forth.

Many students preparing for the AP exam wonder if they need to memorize the 20 amino acids and their structures and whether they are polar, nonpolar, or charged. This is a lot of effort for perhaps one multiple-choice question that you might encounter on the exam. We think that this time would be better spent studying other potential exam questions. If this is of any comfort to you, we have yet to see an AP Biology question that asks something to the effect of "Which of these 5 amino acids is nonpolar?" (*Disclaimer:* This does not mean that it will never happen ☺.) It is more important for you to identify the general structure of an amino acid and know the process of protein synthesis, which we discuss in Chapter 15.

A protein consists of amino acids linked together as shown in Figure 5.8. They are most often much larger than that depicted here. Figure 5.8 is included to enable you to identify a peptide linkage on the exam. Most proteins have many more amino acids in the chain.

The AP exam may expect you to know about the structure of proteins:

Primary structure. The order of the amino acids that make up the protein.

Secondary structure. Three-dimensional arrangement of a protein caused by hydrogen bonding at regular intervals along the polypeptide backbone.

Tertiary structure. Three-dimensional arrangement of a protein caused by interaction among the various R groups of the amino acids involved.

Quaternary structure. The arrangement of separate polypeptide "subunits" into a single protein. Not all proteins have quaternary structure; many consist of a single polypeptide chain.

Proteins with only primary and secondary structure are called *fibrous* proteins. Proteins with only primary, secondary, and tertiary structures are called *globular* proteins. Either fibrous or globular proteins may contain a quaternary structure if there is more than one polypeptide chain.

Enzymes

CT teacher: "The topic of enzymes is full of essay material. Know it well."

Enzymes are proteins that act as organic catalysts and will be encountered often in your review for this exam. **Catalysts** speed up reactions by lowering the energy (activation energy) needed for the reaction to take place, but are not used up in the reaction. The substances that enzymes act on are known as **substrates**.

Enzymes are selective; they interact only with particular substrates. It is the shape of the enzyme that provides the specificity. The part of the enzyme that interacts with the substrate is called the **active site**. The **induced-fit model** of enzyme-substrate interaction

BIG IDEA 4.B.1
The shape of enzymes and their active sites are important to their function.

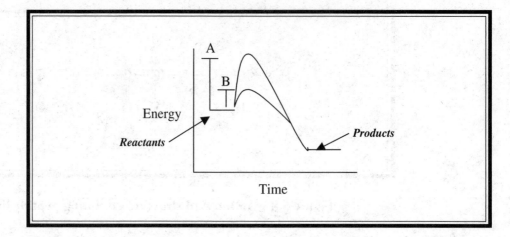

Figure 5.9   Plot showing energy versus time. Height A represents original activation energy; height B represents the lowered activation energy due to the addition of enzyme.

describes the active site of an enzyme as specific for a particular substrate that fits its shape. When the enzyme and substrate bind together, the enzyme is *induced* to alter its shape for a tighter active site–substrate attachment. This tight fit places the substrate in a favorable position to react, speeding up (accelerating) the rate of reaction. After an enzyme interacts with a substrate, converting it into a product, it is free to find and react with another substrate; thus, a small concentration of enzyme can have a major effect on a reaction.

Every enzyme functions best at an optimal temperature and pH. If the pH or temperature strays from those optimal values, the effectiveness of the enzyme will suffer. The effectiveness of an enzyme can be affected by four things:

1. The temperature
2. The pH
3. The concentration of the substrate involved
4. The concentration of the enzyme involved

You should be able to identify the basic components of an activation energy diagram if you encounter one on the AP exam. The important parts are identified in Figure 5.9.

The last enzyme topic to cover is the difference between competitive and noncompetitive inhibition. In **competitive inhibition** (Figure 5.10), an inhibitor molecule resembling

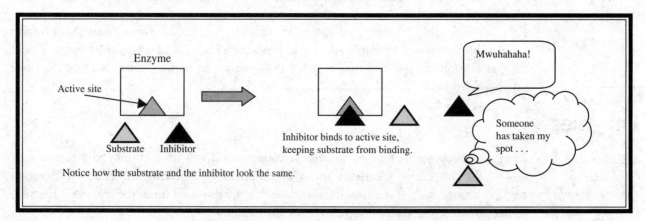

Figure 5.10   Competitive inhibition.

Figure 5.11 Noncompetitive inhibition.

the substrate binds to the active site and physically blocks the substrate from attaching. Competitive inhibition can sometimes be overcome by adding a high concentration of substrate to outcompete the inhibitor. In **noncompetitive inhibition** (Figure 5.11), an inhibitor molecule binds to a different part of the enzyme, causing a change in the shape of the active site so that it can no longer interact with the substrate.

pH: Acids and Bases

The pH scale is used to indicate how acidic or basic a solution is. It ranges from 0 to 14; 7 is neutral. Anything less than 7 is acidic; anything greater than 7 is basic. The pH scale is a logarithmic scale and as a result, a pH of 5 is 10 times more acidic than a pH of 6. Following the same logic, a pH of 4 is 100 times more acidic than a pH of 6. Remember that as the pH of a solution *decreases*, the concentration of hydrogen ions in the solution increases, and vice versa. For the most part, chemical reactions in humans function at or near a neutral pH. The exceptions to this rule are the chemical reactions involving some of the enzymes of the digestive system. (See Chapter 15, Human Physiology.)

Reactions

There are five types of reactions you should know for this exam:

1. *Hydrolysis reaction.* A reaction that breaks down compounds by the addition of H_2O.
2. *Dehydration synthesis reaction.* A reaction in which two compounds are brought together with H_2O released as a product.
3. *Endergonic reaction.* A reaction that requires input of energy to occur.

$$A + B + energy \rightarrow C$$

4. *Exergonic reaction.* A reaction that gives off energy as a product.

$$A + B \rightarrow energy + C$$

5. *Redox reaction.* A reaction involving the transfer of electrons. Such reactions occur along the electron transport chain of the mitochondria during respiration (Chapter 7).

❯ Review Questions

For questions 1–4, please use the following answer choices:

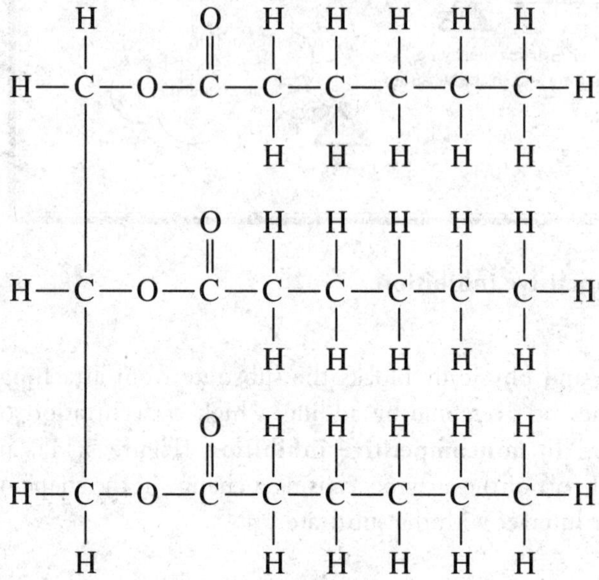

(A)

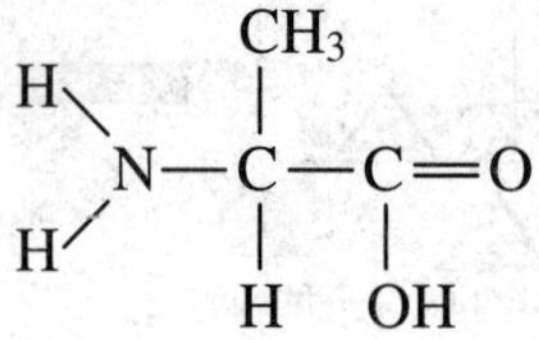

(B)

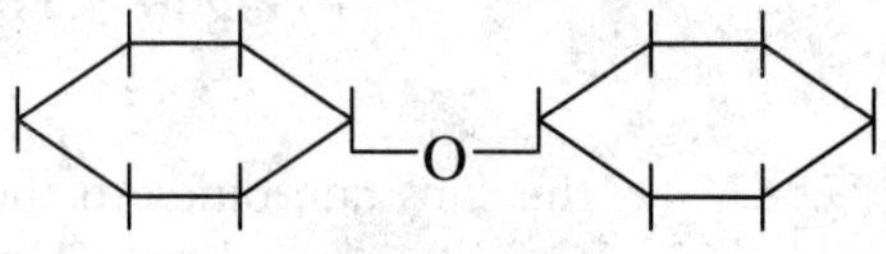

(C)

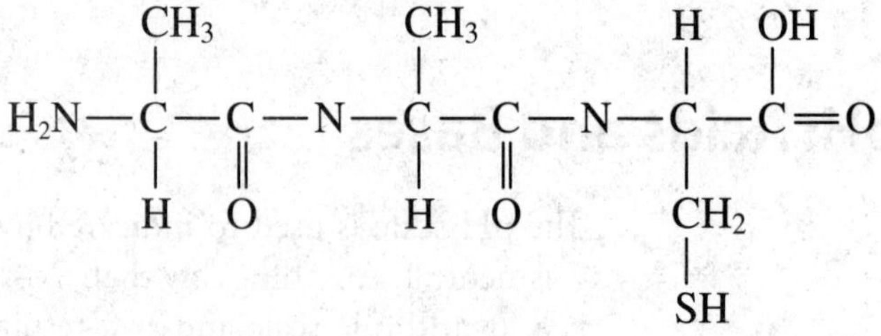

(D)

1. Which of the structures shown above is a polypeptide?

2. Which of these structures is a disaccharide?

3. Which of these structures is a fat?

4. Which of these structures is an amino acid?

5. Which of the following has both a hydrophobic portion and a hydrophilic portion?

   A. Starch
   B. Phospholipids
   C. Proteins
   D. Steroids

6. A solution that has a pH of 2 is how many times more acidic than one with a pH of 5?

   A. 5
   B. 10
   C. 100
   D. 1,000

7. The structure below contains which functional group?

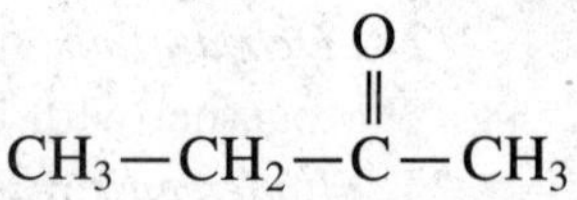

$$CH_3-CH_2-\overset{\overset{\textstyle O}{\|}}{C}-CH_3$$

   A. Aldehyde
   B. Ketone
   C. Amino
   D. Hydroxyl

8. Which of the following will least affect the effectiveness of an enzyme?

   A. Temperature
   B. pH
   C. Concentration of enzyme
   D. Original activation energy of system

9. Which of the following is similar to the process of competitive inhibition?

   A. When you arrive at work in the morning, you are unable to park your car in your (assigned) parking spot because the car of the person who parks next to you has taken up just enough space that you cannot fit your own car in.
   B. When you arrive at work in the morning, you are unable to park your car in your parking spot because someone with a car exactly like yours has already taken your spot, leaving you nowhere to park your car.
   C. As you are about to park your car in your spot at work, a giant bulldozer comes along and smashes your car away from the spot, preventing you from parking your car in your spot.
   D. When you arrive at work in the morning, you are unable to park your car in your parking spot because someone has placed a giant cement block in front of your spot.

10. All the following are carbohydrates except

   A. starch.
   B. glycogen.
   C. chitin.
   D. glycerol.

11. An amino acid contains which of the following functional groups?

   A. Carboxyl group and amino group
   B. Carbonyl group and amino group
   C. Hydroxyl group and amino group
   D. Carboxyl group and hydroxyl group

› Answers and Explanations

1. **D**
2. **C**
3. **A**
4. **B**
5. **B**—A phospholipid has both a hydrophobic portion and a hydrophilic portion. The hydrocarbon portion, or tail, of the phospholipid dislikes water, and the phosphate portion, the head, is hydrophilic.
6. **D**—Because the pH scale is logarithmic, 2 is 1,000 times more acidic than 5.
7. **B**—This functional group is a carbonyl group. The two main types of carbonyl groups are ketones and aldehydes. In this case, it is a ketone because there are carbon chains on either side of the carbon double-bonded to the oxygen.
8. **D**—The four main factors that affect enzyme efficiency are pH, temperature, enzyme concentration, and substrate concentration.
9. **B**—Competitive inhibition is the inhibition of an enzyme–substrate reaction in which the inhibitor resembles the substrate and physically blocks the substrate from attaching to the active site. This parking spot represents the active site, your car is the substrate, and the other car already in the spot is the competitive inhibitor. Examples A and D more closely resemble noncompetitive inhibition.

10. **D**—Glycerol is not a carbohydrate. It is an alcohol. Starch is a carbohydrate stored in plant cells. Glycogen is a carbohydrate stored in animal cells. Chitin is a carbohydrate used by arthropods to construct their exoskeletons. Cellulose is a carbohydrate used by plants to construct their cell walls.

11. A

› Rapid Review

Try to rapidly review the following material:

Organic compounds: contain carbon; examples include lipids, proteins, and carbs (carbohydrates).

Functional groups: amino (NH_2), carbonyl (RCOR), carboxyl (COOH), hydroxyl (OH), phosphate (PO_4), sulfhydryl (SH).

Fat: glycerol + 3 fatty acids.

Saturated fat: bad for you; animals and some plants have it; solidifies at room temperature.

Unsaturated fat: better for you, plants have it; liquifies at room temperature.

Steroids: lipids whose structures resemble chicken-wire fence; include cholesterol and sex hormones.

Phospholipids: glycerol + 2 fatty acids + 1 phosphate group; make up membrane bilayers of cells; have hydrophobic interiors and hydrophilic exteriors.

Carbohydrates: used by cells for energy and structure; monosaccharides (glucose), disaccharides (sucrose, maltose, lactose), storage polysaccharides (starch [plants], glycogen [animals]), structural polysaccharides (chitin [fungi], cellulose [arthropods]).

Proteins: made with the help of ribosomes out of amino acids; serve many functions (e.g., transport, enzymes, cell signals, receptor molecules, structural components, and channels).

Enzymes: catalytic proteins that react in an induced-fit fashion with substrates to speed up the rate of reactions by lowering the activation energy; effectiveness is affected by changes in pH, temperature, and substrate and enzyme concentrations.

Competitive inhibition: inhibitor resembles substrate and binds to active site.

Noncompetitive inhibition: inhibitor binds elsewhere on enzyme; alters active site so that substrate cannot bind.

pH: logarithmic scale <7 acidic, 7 neutral, >7 basic (alkaline); pH 4 is 10 times more acidic than pH 5.

Reaction types:

  Hydrolysis reaction: breaks down compounds by adding water.

  Dehydration reaction: two components brought together, producing H_2O.

  Endergonic reaction: reaction that requires input of energy.

  Exergonic reaction: reaction that gives off energy.

  Redox reaction: electron transfer reactions.

CHAPTER 6

Cells

IN THIS CHAPTER

Summary: This chapter discusses the different types of cells (eukaryotic and pro-karyotic) and the important organelles, structures, and transport mechanisms that power these cells.

Key Ideas

○ Prokaryotic cells are simple cells with no nuclei or organelles.
○ Animal cells do not contain cell walls or chloroplasts and have small vacuoles.
○ Plant cells do not have centrioles.
○ The fluid mosaic model states that a cell membrane consists of a phospholipid bilayer with proteins of various lengths and sizes interspersed with cholesterol among the phospholipids.
○ Passive transport is the movement of a particle across a selectively permeable membrane down its concentration gradient (e.g., diffusion, osmosis).
○ Active transport is the movement of a particle across a selectively permeable membrane against its concentration gradient (e.g., sodium-potassium pump).

Introduction

A cell is defined as a small room, sometimes a prison room, usually designed for only one person (but usually housing two or more inmates, except for solitary-confinement cells). It is a place for rehabilitation—whoops! We're looking at the wrong notes here. Sorry, let's start again. A cell is the basic unit of life (that's more like it), discovered in the seventeenth century by Robert Hooke. There are two major divisions of cells: prokaryotic and eukaryotic.

This chapter starts with a discussion of these two cell types, followed by an examination of the organelles found in cells. We conclude with a look at the fluid mosaic model of the cell membrane and a discussion of the different types of cell transport: diffusion, facilitated diffusion, osmosis, active transport, endocytosis, and exocytosis.

Types of Cells

The **prokaryotic** cell is a *simple* cell. It has no nucleus, and no membrane-bound organelles. The genetic material of a prokaryotic cell is found in a region of the cell known as the **nucleoid.** Bacteria are a fine example of prokaryotic cells and divide by a process known as *binary fission*; they duplicate their genetic material, divide in half, and produce two identical daughter cells. Prokaryotic cells are found only in the kingdom Monera (bacteria group).

Steve (12th grade): "Five questions on my test dealt with organelle function—know them."

The **eukaryotic** cell is much more complex. It contains a nucleus, which functions as the control center of the cell, directing DNA replication, transcription, and cell growth. Eukaryotic organisms may be unicellular or multicellular. One of the key features of eukaryotic cells is the presence of membrane-bound organelles, each with its own duties. Two prominent members of the "Eukaryote Club" are animal and plant cells; the differences between these types of cells are discussed in the next section.

Organelles

BIG IDEA 1.B.2
Cell structure is an example of a widely conserved feature.

You should familiarize yourselves with approximately a dozen organelles and cell structures before taking the AP Biology exam:

Prokaryotic Organelles

You should be familiar with the following structures:

Plasma membrane. This is a selective barrier around a cell composed of a double layer of phospholipids. Part of this selectivity is due to the many proteins that either rest on the exterior of the membrane or are embedded in the membrane of the cell. Each membrane has a different combination of lipids, proteins, and carbohydrates that provide it with its unique characteristics.

Cell wall. This is a wall or barrier that functions to shape and protect cells. This is present in all prokaryotes.

Ribosomes. These function as the host organelle for protein synthesis in the cell. They are found in the cytoplasm of cells and are composed of a large unit and a small subunit.

Eukaryotic Organelles

You should be familiar with the following structures:

BIG IDEA 2.B.3
Eukaryotic cells have organelles to partition such cells into specialized regions.

Ribosomes. As in prokaryotes, eukaryotic ribosomes serve as the host organelles for protein synthesis. Eukaryotes have *bound* ribosomes, which are attached to endoplasmic reticula and form proteins that tend to be exported from the cell or sent to the membrane. There are also *free* ribosomes, which exist freely in the cytoplasm and produce proteins that remain in the cytoplasm of the cell. Eukaryotic ribosomes are built in a structure called the **nucleolus.**

BIG IDEA 4.A.2
The structure and function of organelles provide essential processes for the cell.

Smooth endoplasmic reticulum. This is a membrane-bound organelle involved in lipid synthesis, detoxification, and carbohydrate metabolism. Liver cells contain a lot of **smooth endoplasmic reticulum** (SER) because they host a lot of carbohydrate metabolism (glycolysis). It is given the name "smooth" endoplasmic reticulum because there are no ribosomes on its cytoplasmic surface. The liver contains much SER for another reason—it is the site of alcohol detoxification.

Rough endoplasmic reticulum. This membrane-bound organelle is termed "rough" because of the presence of ribosomes on the cytoplasmic surface of the cell. The proteins produced by this organelle are often secreted by the cell and carried by vesicles to the **Golgi apparatus** for further modification.

Golgi apparatus. Proteins, lipids, and other macromolecules are sent to the Golgi to be modified by the addition of sugars and other molecules to form **glycoproteins.** The products are then sent in vesicles (escape pods that bud off the edge of the Golgi) to other parts of the cell, directed by the particular changes made by the Golgi. We think of the Golgi apparatus as the post office of the cell—packages are dropped off by customers, and the Golgi adds the appropriate postage and zip code to make sure that the packages reach proper destinations in the cell.

Mitochondria. These are double-membraned organelles that specialize in the production of ATP. The innermost portion of the mitochondrion is called the *matrix,* and the folds created by the inner of the two membranes are called *cristae.* The mitochondria are the host organelles for the Krebs cycle (matrix) and oxidative phosphorylation (cristae) of respiration, which we discuss in Chapter 7. We think of the mitochondria as the power plants of the cell.

Lysosome. This is a membrane-bound organelle that specializes in digestion. It contains enzymes that break down (hydrolyze) proteins, lipids, nucleic acids, and carbohydrates. This organelle is the stomach of the cell. Absence of a particular lysosomal hydrolytic enzyme can lead to a variety of diseases known as **storage diseases.** An example of this is **Tay-Sachs disease** (discussed in Chapter 10), in which an enzyme used to digest lipids is absent, leading to excessive accumulation of lipids in the brain. Lysosomes are often referred to as "suicide sacs" of the cell. Cells that are no longer needed are often destroyed in these sacs. An example of this process involves the cells of the tail of a tadpole, which are digested as a tadpole changes into a frog.

Nucleus. This is the control center of the cell. In eukaryotic cells, this is the storage site of genetic material (DNA). It is the site of replication, transcription, and posttranscriptional modification of RNA. It also contains the nucleolus, the site of ribosome synthesis.

Vacuole. This is a storage organelle that acts as a vault. Vacuoles are quite large in plant cells but small in animal cells.

Peroxisomes. These are organelles containing enzymes that produce hydrogen peroxide as a by-product while performing various functions, such as breakdown of fatty acids and detoxification of alcohol in the liver. Peroxisomes also contain an enzyme that converts the toxic hydrogen peroxide by-product of these reactions into cell-friendly water.

Chloroplast. This is the site of photosynthesis and energy production in plant cells. Chloroplasts contain many pigments, which provide leaves with their color. Chloroplasts are divided into an inner portion and an outer portion. The inner fluid portion is called the **stroma,** which is surrounded by two outer membranes. Winding through the stroma is an inner membrane called the **thylakoid membrane system,** where the light-dependent reactions of photosynthesis occur. The light-independent (dark) reactions occur in the stroma.

Cytoskeleton. The skeleton of cells consists of three types of fibers that provide support, shape, and mobility to cells: microtubules, microfilaments, and intermediate filaments. **Microtubules** are constructed from tubulin and have a lead role in the separation of cells during cell division. Microtubules are also important components of cilia and flagella, which are structures that aid the movement of particles (Chapter 19). **Microfilaments,** constructed from actin, play a big part in muscular contraction. **Intermediate filaments** are constructed from a class of proteins called *keratins* and are thought to function as reinforcement for the shape and position of organelles in the cell.

Remember me!

Of the structures listed above, animal cells contain *all except* cell walls and chloroplasts, and their vacuoles are small. Plant cells contain *all* the structures listed above, and their vacuoles are large. Animal cells have centrioles (cell division structure); plant cells *do not*!

Cell Membranes: Fluid Mosaic Model

As discussed above and in Chapter 5, a cell membrane is a selective barrier surrounding a cell that has a phospholipid bilayer as its major structural component. Remember that the outer portion of the bilayer contains the hydrophilic (water-loving) head of the phospholipid, while the inner portion is composed of the hydrophobic (water-fearing) tail of the phospholipid (Figure 6.1).

The **fluid mosaic model** is the most accepted model for the arrangement of membranes. It states that the membrane consists of a phospholipid bilayer with proteins of various lengths and sizes interspersed with cholesterol among the phospholipids. These proteins perform various functions depending on their location within the membrane.

The fluid mosaic model consists of **integral proteins,** which are implanted within the bilayer and can extend partway or all the way across the membrane, and **peripheral proteins,** such as receptor proteins, which are not implanted in the bilayer and are often attached to integral proteins of the membrane. These proteins have various functions in cells. A protein that stretches across the membrane can function as a channel to assist the passage of desired molecules into the cell. Proteins on the exterior of a membrane with

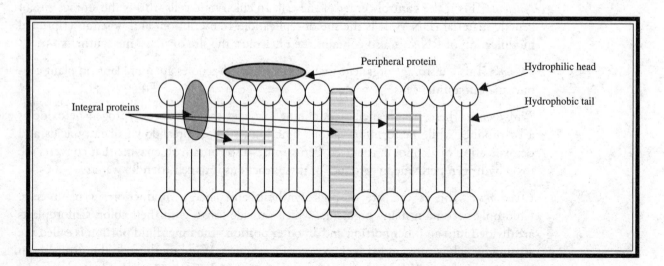

Figure 6.1   Cross-section of a cell membrane showing phospholipid bilayer.

binding sites can act as receptors that allow the cell to respond to external signals such as hormones. Proteins embedded in the membrane can also function as enzymes, increasing the rate of cellular reactions.

The cell membrane is "selectively" permeable, meaning that it allows some molecules and other substances through, while others are not permitted to pass. The membrane is like a bouncer at a popular nightclub. What determines the selectivity of the membrane? One factor is the size of the substance, and the other is the charge. The bouncer lets small, uncharged polar substances and hydrophobic substances such as lipids through the membrane, but larger uncharged polar substances (such as glucose) and charged ions (such as sodium) cannot pass through. The other factor determining what is allowed to pass through the membrane is the particular arrangement of proteins in the lipid bilayer. Different proteins in different arrangements allow different molecules to pass through.

Types of Cell Transport

There are six basic types of cell transport:

1. **Diffusion:** the movement of molecules down their concentration gradient without the use of energy. It is a *passive* process during which substances move from a region of higher concentration to a region of lower concentration. The rate of diffusion of substances varies from membrane to membrane because of different selective permeabilities.

BIG IDEA 2.B.2
Cells maintain their internal environment by transporting materials across their membranes.

2. **Osmosis:** the *passive* diffusion of water down its concentration gradient across selectively permeable membranes. Water moves from a region of *high* water concentration to a region of *low* water concentration. Thinking about osmosis another way, water will flow from a region with a *lower* solute concentration (hypotonic) to a region with a *higher* solute concentration (hypertonic). This process does not require the input of energy. For example, visualize two regions—one with 10 particles of sodium per liter of water; the other with 15. Osmosis would drive water from the region with 10 particles of sodium toward the region with 15 particles of sodium.

3. **Facilitated diffusion:** the diffusion of particles across a selectively permeable membrane with the assistance of the membrane's transport proteins. These proteins will not bring any old molecule looking for a free pass into the cell; they are specific in what they will carry and have binding sites designed for molecules of interest. Like diffusion and osmosis, this process does not require the input of energy.

4. **Active transport:** the movement of a particle across a selectively permeable membrane *against* its concentration gradient (from low concentration to high). This movement requires the input of energy, which is why it is termed "active" transport. As is often the case in cells, adenosine triphosphate (ATP) is called on to provide the energy for this reactive process. These active-transport systems are vital to the ability of cells to maintain particular concentrations of substances despite environmental concentrations. For example, cells have a very high concentration of potassium and a very low concentration of sodium. Diffusion would like to move sodium in and potassium out to equalize the concentrations. The all-important **sodium-potassium pump** actively moves potassium *into* the cell and sodium *out of* the cell against their respective concentration gradients to maintain appropriate levels inside the cell. This is the major pump in animal cells.

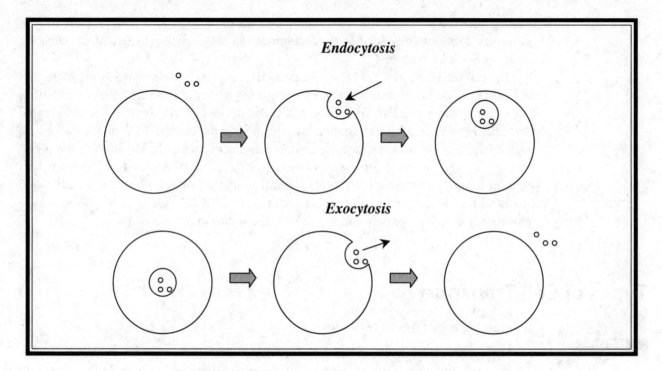

Figure 6.2 Endocytosis and exocytosis.

5. **Endocytosis:** a process in which substances are brought into cells by the enclosure of the substance into a membrane-created vesicle that surrounds the substance and escorts it into the cell (Figure 6.2). This process is used by immune cells called **phagocytes** to engulf and eliminate foreign invaders.

6. **Exocytosis:** a process in which substances are exported out of the cell (the reverse of endocytosis). A vesicle again escorts the substance to the plasma membrane, causes it to fuse with the membrane, and ejects the contents of the substance outside the cell (Figure 6.2). In exocytosis, the vesicle functions like the trash chute of the cell.

› Review Questions

For questions 1–4, please use the following answer choices:

   A. Cell wall
   B. Mitochondrion
   C. Ribosome
   D. Lysosome

1. This organelle is present in plant cells, but not animal cells.

2. Absence of enzymes from this organelle can lead to storage diseases such as Tay-Sachs disease.

3. This organelle is the host for the Krebs cycle and oxidative phosphorylation of respiration.

4. This organelle is synthesized in the nucleolus of the cell.

5. Which of the following best describes the fluid mosaic model of membranes?

   A. The membrane consists of a phospholipid bilayer with proteins of various lengths and sizes located on the exterior portions of the membrane.
   B. The membrane consists of a phospholipid bilayer with proteins of various lengths and sizes located in the interior of the membrane.
   C. The membrane is composed of a phospholipid bilayer with proteins of uniform lengths and sizes located in the interior of the membrane.
   D. The membrane contains a phospholipid bilayer with proteins of various lengths and sizes interspersed among the phospholipids.

6. Which of the following types of cell transport requires energy?

   A. The movement of a particle across a selectively permeable membrane down its concentration gradient
   B. The movement of a particle across a selectively permeable membrane against its concentration gradient
   C. The movement of water down its concentration gradient across selectively permeable membranes
   D. The movement of a sodium ion from an area of higher concentration to an area of lower concentration

7. Which of the following structures is present in prokaryotic cells?

   A. Nucleus
   B. Mitochondria
   C. Cell wall
   D. Golgi apparatus

8. Which of the following represents an *incorrect* description of an organelle's function?

   A. *Chloroplast:* the site of photosynthesis and energy production in plant cells
   B. *Peroxisome:* organelle that produces hydrogen peroxide as a by-product of reactions involved in the breakdown of fatty acids, and detoxification of alcohol in the liver
   C. *Golgi apparatus:* structure to which proteins, lipids, and other macromolecules are sent to be modified by the addition of sugars and other molecules to form glycoproteins
   D. *Rough endoplasmic reticulum:* membrane-bound organelle lacking ribosomes on its cytoplasmic surface, involved in lipid synthesis, detoxification, and carbohydrate metabolism

9. The destruction of which of the following would most cripple a cell's ability to undergo cell division?

   A. Microfilaments
   B. Intermediate filaments
   C. Microtubules
   D. Actin fibers

10. Which of the following can easily diffuse across a selectively permeable membrane?

    A. Na^+
    B. Glucose
    C. Large uncharged polar molecules
    D. Lipids

› Answers and Explanations

1. **A**—Cell walls exist in plant cells and prokaryotic cells, but not animal cells. They function to shape and protect cells.

2. **D**—The lysosome acts like the stomach of the cell. It contains enzymes that break down proteins, lipids, nucleic acids, and carbohydrates. Absence of these enzymes can lead to storage disorders such as Tay-Sachs disease.

3. **B**—The mitochondrion is the power plant of the cell. This organelle specializes in the production of ATP and hosts the Krebs cycle and oxidative phosphorylation.

4. **C**—The ribosome is an organelle made in the nucleolus that serves as the host for protein synthesis in the cell. It is found in both prokaryotes and eukaryotes.

5. **D**—The fluid mosaic model says that proteins can extend all the way through the phospholipid bilayer of the membrane, and that these proteins are of various sizes and lengths.

6. **B**—Answer choice B is the definition of active transport, which requires the input of energy. Simple diffusion (answer choices A and D) and osmosis (answer choice C) are all passive processes that do not require energy input.

7. **C**—Prokaryotes do not contain many organelles, but they do contain cell walls.

8. **D**—This is the description of the *smooth* endoplasmic reticulum. We know that this is a tricky question, but we wanted you to review the distinction between the two types of endoplasmic reticulum.

9. **C**—Microtubules play an enormous role in cell division. They make up the spindle apparatus that works to pull apart the cells during mitosis (Chapter 9). A loss of microtubules would cripple the cell division process. Actin fibers (answer choice D) are the building blocks of microfilaments (answer choice A), which are involved in muscular contraction. Keratin fibers are the building blocks of intermediate filaments (answer choice B), which function as reinforcement for the shape and position of organelles in the cell.

10. **D**—Lipids are the only substances listed that are able to freely diffuse across selectively permeable membranes.

› Rapid Review

Try to rapidly review the materials presented in the following table and list:

ORGANELLE	PROKARYOTES	ANIMAL CELLS EUKARYOTES	PLANT CELLS EUKARYOTES	FUNCTION
Cell wall	+	−	+	Protects and shapes the cell
Plasma membrane	+	+	+	Regulates what substances enter and leave a cell
Ribosome	+	+	+	Host for protein synthesis; formed in nucleolus
Smooth ER*	−	+	+	Lipid synthesis, detoxification, carbohydrate metabolism; no ribosomes on cytoplasmic surface
Rough ER*	−	+	+	Synthesizes proteins to secrete or send to plasma membrane; contains ribosomes on cytoplasmic surface
Golgi	−	+	+	Modifies lipids, proteins, etc., and sends them to other sites in the cell
Mitochondria	−	+	+	Power plant of cell; hosts major energy-producing steps of respiration
Lysosome	−	+	−	Contains enzymes that digest organic compounds; serves as cell's stomach
Nucleus	−	+	+	Control center of cell; host for transcription, replication, and DNA
Peroxisome	−	+	+	Breakdown of fatty acids, detoxification of alcohol
Chloroplast	−	−	+	Site of photosynthesis in plants
Cytoskeleton	−	+	+	Skeleton of cell; consists of microtubules (cell division, cilia, flagella), microfilaments (muscles), and intermediate filaments (reinforcing position of organelles)
Vacuole	−	+, small	+, large	Storage vault of cells
Centrioles	−	+	−	Part of microtubule separation apparatus that assists cell division in animal cells

*Endoplasmic reticulum

Fluid mosaic model: plasma membrane is a selectively permeable phospholipid bilayer with proteins of various lengths and sizes interspersed with cholesterol among the phospholipids.

Integral proteins: proteins implanted within lipid bilayer of plasma membrane.

Peripheral proteins: proteins attached to exterior of membrane.

Diffusion: passive movement of substances down their concentration gradient (from high to low concentrations).

Osmosis: passive movement of water from the side of low solute concentration to the side of high solute concentration (hypotonic to hypertonic).

Facilitated diffusion: assisted transport of particles across membrane (no energy input needed).

Active transport: movement of substances against concentration gradient (low to high concentrations; requires energy input).

Endocytosis: phagocytosis of particles into a cell through the use of vesicles.

Exocytosis: process by which particles are ejected from the cell, similar to movement in a trash chute.

CHAPTER 7

Respiration

IN THIS CHAPTER

Summary: This chapter covers the basics behind the energy-creation process known as respiration. This chapter also teaches you the difference between aerobic and anaerobic respiration and takes you through the steps that convert a glucose molecule into ATP.

Key Ideas

- ✪ Aerobic respiration: glycolysis → Krebs cycle → oxidative phosphorylation → 36 ATP.
- ✪ Anaerobic respiration: glycolysis → regenerate NAD^+ → much less ATP.
- ✪ Oxidative phosphorylation results in the production of large amounts of ATP from NADH and $FADH_2$.
- ✪ Chemiosmosis is the coupling of the movement of electrons down the electron transport chain with the formation of ATP using the driving force provided by the proton gradient.

Introduction

BIG IDEA 2.A.1

All living things require constant input of energy.

In this chapter, we explore how cells obtain energy. It is important that you do not get lost or buried in the details. You should finish this chapter with an understanding of the basic process. The AP Biology exam will not ask you to identify by name the enzyme that catalyzes the third step of glycolysis, nor will it require you to name the fourth molecule in the Krebs cycle. But it *will* ask you questions that require an understanding of the respiration process.

There are two major categories of respiration: **aerobic** and **anaerobic.** Aerobic respiration occurs in the presence of oxygen, while anaerobic respiration occurs in situations where oxygen is not available. Aerobic respiration involves three stages: glycolysis, the Krebs cycle, and oxidative phosphorylation. Anaerobic respiration, sometimes referred to as *fermentation,* also begins with glycolysis, and concludes with the formation of NAD$^+$.

Aerobic Respiration

Glycolysis

BIG IDEA 2.A.2

Heterotrophs capture free energy present in the food they eat through cellular respiration.

Glycolysis occurs in the cytoplasm of cells and is the beginning pathway for both aerobic and anaerobic respiration. During glycolysis, a glucose molecule is broken down through a series of reactions into two molecules of pyruvate. It is important to remember that oxygen plays no role in glycolysis. This reaction can occur in oxygen-rich and oxygen-poor environments. However, when in an environment lacking oxygen, glycolysis slows because the cells run out (become depleted) of NAD$^+$. For reasons we will discuss later, a lack of oxygen prevents oxidative phosphorylation from occurring, causing a buildup of NADH in the cells. This buildup causes a shortage of NAD$^+$. This is bad for glycolysis because it requires NAD$^+$ to function. Fermentation is the solution to this problem—it takes the excess NADH that builds up and converts it back to NAD$^+$ so that glycolysis can continue. More to come on fermentation later . . . be patient. ☺

To reiterate, the AP Biology exam will not require you to memorize the various steps of respiration. Your time is better spent studying the broad explanation of respiration, to understand the basic process, and become comfortable with respiration as a whole. Major concepts are the key. We will explain the specific steps of glycolysis because they will help you understand the big picture—but do not memorize them all. Save the space for other facts you have to know from other chapters of this book.

Examine Figure 7.1, which illustrates the general layout of glycolysis. The beginning steps of glycolysis require energy input. The first step adds a phosphate to a molecule of glucose with the assistance of an ATP molecule to produce *glucose-6-phosphate* (G6P). The newly formed G6P rearranges to form a molecule named *fructose-6-phosphate* (F6P). Another molecule of ATP is required for the next step, which adds another phosphate group to produce fructose 1,6-biphosphate. Already, glycolysis has used two of the ATP molecules that it is trying to produce—seems stupid . . . but be patient . . . the genius has yet to show its face. F6P splits into two 3-carbon-long fragments known as **PGAL** (glyceraldehyde phosphate). With the formation of PGAL, the energy-producing portion of glycolysis begins. Each PGAL molecule takes on an inorganic phosphate from the cytoplasm to produce 1,3-diphosphoglycerate. During this reaction, each PGAL gives up two electrons and a hydrogen to molecules of NAD$^+$ to form the all-important NADH molecules. The next step is a big one, as it leads to the production of the first ATP molecule in the process of respiration—the 1,3-diphosphoglycerate molecules donate one of their two phosphates to molecules of ADP to produce ATP and 3-phosphoglycerate (3PG). You'll notice that there are *two* ATP molecules formed here because before this step, the single molecule of glucose divided into *two* 3-carbon fragments. After 3PG rearranges to form 2-phosphoglycerate, phosphoenolpyruvate (PEP) is formed, which donates a phosphate group to molecules of ADP to form another pair of ATP molecules and pyruvate. This is the final step of glycolysis. In total, two molecules each of ATP, NADH, and pyruvate are formed during this process. Glycolysis produces the same result under anaerobic conditions as it does under aerobic conditions: two ATP molecules. If oxygen is present, more ATP is later made by oxidative phosphorylation.

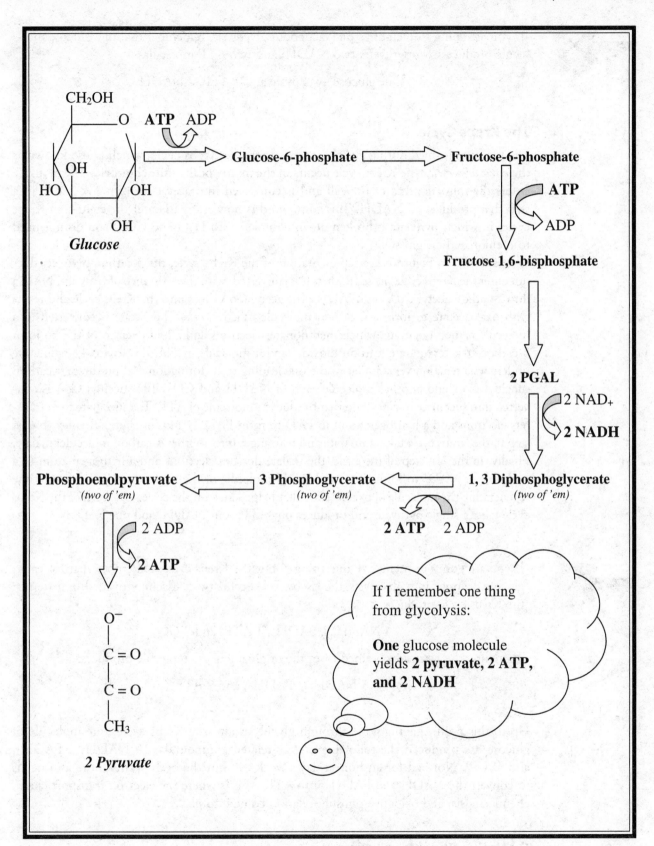

Figure 7.1   Glycolysis.

If you are going to memorize one fact about glycolysis, remember that one glucose molecule produces two pyruvate, two NADH, and two ATP molecules.

One glucose → 2 pyruvate, 2 ATP, 2 NADH

The Krebs Cycle

The pyruvate formed during glycolysis next enters the **Krebs cycle,** which is also known as the *citric acid cycle.* The Krebs cycle occurs in the matrix of the **mitochondria.** The pyruvate enters the mitochondria of the cell and is converted into acetyl coenzyme A (CoA) in a step that produces an NADH. This compound is now ready to enter the eight-step Krebs cycle, in which pyruvate is broken down completely to H_2O and CO_2. You do not need to memorize the eight steps.

As shown in Figure 7.2, a representation of the Krebs cycle, the 3-carbon pyruvate does not enter the Krebs cycle per se. Rather, it is converted, with the assistance of CoA and NAD^+, into 2-carbon acetyl CoA and NADH. The acetyl CoA dives into the Krebs cycle and reacts with oxaloacetate to form a 6-carbon molecule called *citrate.* The citrate is converted to a molecule named isocitrate, which then donates electrons and a hydrogen to NAD^+ to form 5-carbon α-ketoglutarate, carbon dioxide, and a molecule of NADH. The α-ketoglutarate undergoes a reaction very similar to the one leading to its formation and produces 4-carbon succinyl CoA and another molecule each of NADH and CO_2. The succinyl CoA is converted into succinate in a reaction that produces a molecule of ATP. The succinate then transfers electrons and a hydrogen atom to FAD to form $FADH_2$ and fumarate. The next-to-last step in the Krebs cycle takes fumarate and rearranges it to another 4-carbon molecule: malate. Finally, in the last step of the cycle, the malate donates electrons and a hydrogen atom to a molecule of NAD^+ to form the final NADH molecule of the Krebs cycle, at the same time regenerating the molecule of oxaloacetate that helped kick off the cycle. One turn of the Krebs cycle takes a single pyruvate and produces one ATP, four NADH, and one $FADH_2$.

If you are going to memorize one thing about the Krebs cycle, remember that for each glucose dropped into glycolysis, the Krebs cycle occurs twice. Each pyruvate dropped into the Krebs cycle produces

4 NADH, 1 $FADH_2$, 1 ATP, and 2 CO_2

Therefore, the *pyruvate* obtained from the original glucose molecule produces:

8 NADH, 2 $FADH_2$, and 2 ATP

Up to this point, having gone through glycolysis and the Krebs cycle, one molecule of glucose has produced the following energy-related compounds: 10 NADH, 2 $FADH_2$, and 4 ATP. Not bad for an honest day's work . . . but the body wants more and needs to convert the NADH and $FADH_2$ into ATP. This is where the electron transport chain, chemiosmosis, and oxidative phosphorylation come into play.

Oxidative Phosphorylation

After the Krebs cycle comes the largest energy-producing step of them all: **oxidative phosphorylation.** During this aerobic process, the NADH and $FADH_2$ produced during the first two stages of respiration are used to create ATP. Each NADH leads to

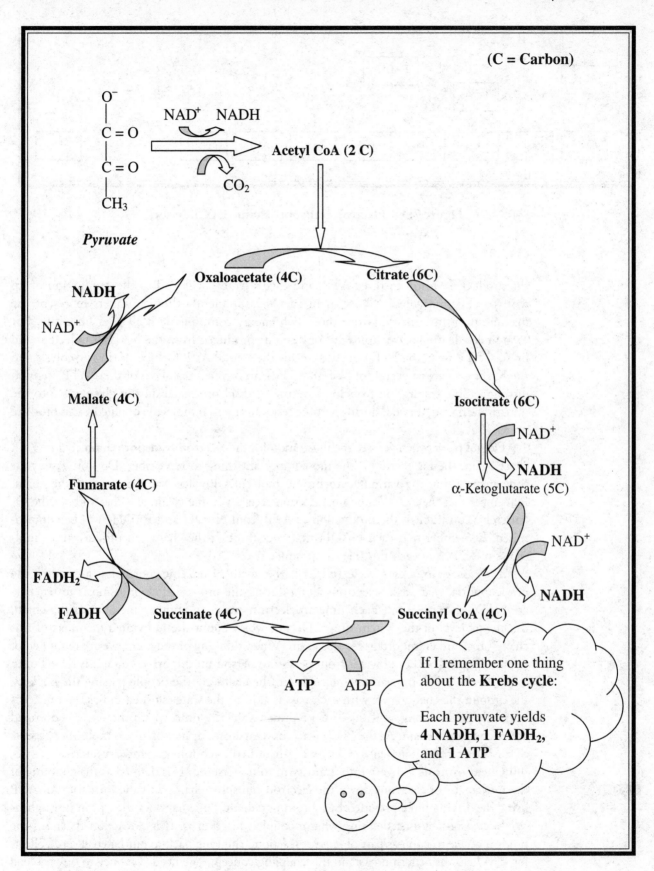

Figure 7.2   The Krebs cycle.

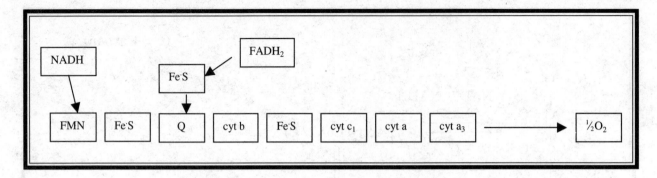

Figure 7.3   Electron transport chain (ETC).

the production of up to three ATP, and each $FADH_2$ will lead to the production of up to two ATP molecules. This is an inexact measurement—those numbers represent the maximum output possible from those two energy components if all goes smoothly. For each molecule of glucose, up to 30 ATP can be produced from the NADH molecules and up to 4 ATP from the $FADH_2$. Add to this the 4 total ATP formed during glycolysis and the Krebs cycle for a grand total of 38 ATP from *each glucose.* Two of these ATP are used during aerobic respiration to help move the NADH produced during glycolysis into the mitochondria. All totaled, during aerobic respiration, each molecule of glucose can produce up to 36 ATP.

Do not panic when you see the illustration for the **electron transport chain** (Figure 7.3). Once again, the big picture is the most important thing to remember. Do not waste your time memorizing the various cytochrome molecules involved in the steps of the chain. Remember that the ½ O_2 is the final electron acceptor in the chain, and that without the O_2 (anaerobic conditions), the production of ATP from NADH and $FADH_2$ will be compromised. Remember that each NADH that goes through the chain can produce three molecules of ATP, and each $FADH_2$ can produce two.

The *electron transport chain* (ETC) is the chain of enzyme molecules, located in the mitochondria, that passes electrons along during the process of chemiosmosis to regenerate NAD^+ to form ATP. Each time an electron passes to another member of the chain, the energy level of the system drops. Do not worry about the individual members of this chain—they are unimportant for this exam. When thinking of the ETC, we are reminded of the passing of a bucket of water from person to person until it arrives at and is tossed onto a fire. In the ETC, the various molecules in the chain are the people passing the buckets; the drop in the energy level with each pass is akin to the water sloshed out as the bucket is hurriedly passed along, and the ½ O_2 represents the fire onto which the water is dumped at the end of the chain. As the ½ O_2 (each oxygen atom, or half of an O_2 molecule) accepts a pair of electrons, it actually picks up a pair of hydrogen ions to *produce* water.

Chemiosmosis is a very important term to understand. It is defined as the coupling of the movement of electrons down the electron transport chain with the formation of ATP using the driving force provided by a proton gradient. So, what does that mean in English? Well, let's start by first defining what a coupled reaction is. It is a reaction that uses the product of *one* reaction as part of *another* reaction. Thinking back to our baseball card collecting days helps us better understand this coupling concept. We needed money to buy baseball cards. We would babysit or do yardwork for our neighbors and use that money to buy cards. We coupled the money-making reaction of hard labor to the money-spending reaction of buying baseball cards.

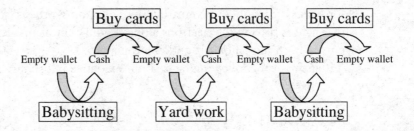

Let's look more closely at the reactions that are coupled in chemiosmosis. If you look at Figure 7.4a, a crude representation of a mitochondrion, you will find the ETC embedded within the inner mitochondrial membrane. As some of the molecules in the chain accept and then pass on electrons, they pump hydrogen ions into the space between the inner and outer membranes of the mitochondria (Figure 7.4b). This creates a proton gradient that drives the production of ATP. The difference in hydrogen concentration on the two sides of the membrane causes the protons to flow back into the matrix of the mitochondria through ATP synthase channels (Figure 7.4c). **ATP synthase** is an enzyme that uses the flow of hydrogens to drive the phosphorylation of an ADP molecule to produce ATP. This reaction completes the process of oxidative phosphorylation and chemiosmosis. The proton gradient created by the movement of electrons from molecule to molecule has been used to form the ATP that this process is designed to produce. In other words, the formation of ATP has been coupled to the movement of electrons and protons.

Chemiosmosis is not oxidative phosphorylation per se; rather, it is a major *part* of oxidative phosphorylation. An important fact we want you to take out of this chapter is that chemiosmosis is not unique to the mitochondria. It is the same process that occurs in the

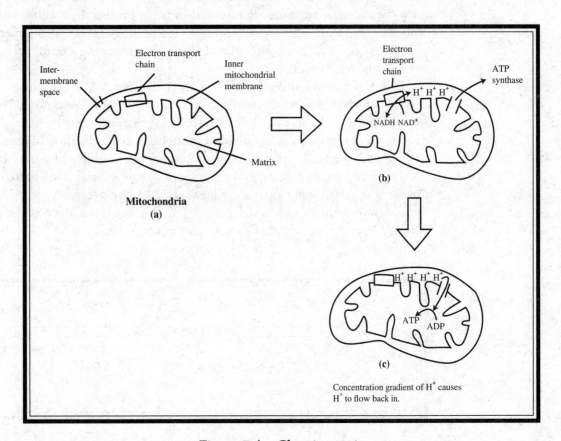

Figure 7.4   Chemiosmosis.

chloroplasts during the ATP-creating steps of photosynthesis (see Chapter 8). The difference is that light is driving the electrons along the ETC in plants. Remember that chemiosmosis occurs in both mitochondria and chloroplasts.

Remember the following facts about oxidative phosphorylation (Ox-phos):

1. Each NADH → 3 ATP.
2. Each FADH$_2$ → 2 ATP.
3. ½ O$_2$ is the final electron acceptor of the electron transport chain, and the chain will not function in the absence of oxygen.
4. Ox-phos serves the important function of regenerating NAD$^+$ so that glycolysis and the Krebs cycle can continue.
5. Chemiosmosis occurs in photosynthesis as well as respiration.

Anaerobic Respiration

Anaerobic respiration, or *fermentation,* occurs when oxygen is unavailable or cannot be used by the organism. As in aerobic respiration, glycolysis occurs and pyruvate is produced. The pyruvate enters the Krebs cycle, producing NADH, FADH$_2$, and some ATP. The problem arises in the ETC—because there is no oxygen available, the electrons do not pass down the chain to the final electron acceptor, causing a buildup of NADH in the system. This buildup of NADH means that the NAD$^+$ normally regenerated during oxidative phosphorylation is not produced, and this creates an NAD$^+$ shortage. This is a problem, because in order for glycolysis to proceed to the pyruvate stage, it needs NAD$^+$ to help perform the necessary reactions. **Fermentation** is the process that begins with glycolysis and ends when NAD$^+$ is regenerated. A glucose molecule that enters the fermentation pathway produces two net ATP per molecule of glucose, representing a tremendous decline in the efficiency of ATP production.

Under aerobic conditions, NAD$^+$ is recycled from NADH by the movement of electrons down the electron transport chain. Under anaerobic conditions, NAD$^+$ is recycled from NADH by the movement of electrons to pyruvate, namely, fermentation. The two main types of fermentation are **alcohol fermentation** and **lactic acid fermentation.** Refer to Figures 7.5 and 7.6 for the representations of the different forms of fermentation. Alcohol fermentation (Figure 7.5) occurs in fungi, yeast, and some bacteria. The first step involves the conversion of pyruvate into two 2-carbon acetaldehyde molecules. Then, in the all-important step of alcohol fermentation, the acetaldehyde molecules are converted to ethanol, regenerating two NAD$^+$ molecules in the process.

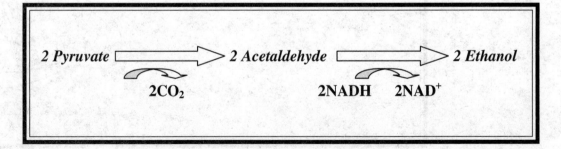

Figure 7.5   Alcohol fermentation.

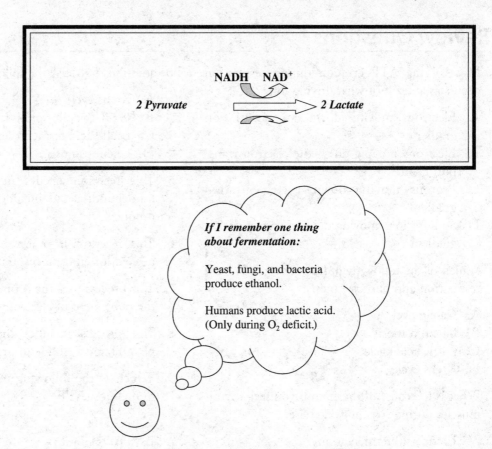

Figure 7.6   Lactic acid fermentation.

Lactic acid fermentation (Figure 7.6) occurs in human and animal muscle cells when oxygen is not available. This is a simpler process than alcoholic fermentation—the pyruvate is directly reduced to lactate (also known as lactic acid) by NADH to regenerate the NAD^+ needed for the resumption of glycolysis. Have you ever had a cramp during exercise? The pain you felt was the result of lactic acid fermentation. Your muscle was deprived of the necessary amount of oxygen to continue glycolysis, and it switched over to fermentation. The pain from the cramp came from the acidity in the muscle.

› Review Questions

1. Most of the ATP creation during respiration occurs as a result of what driving force?

   A. Electrons moving down a concentration gradient
   B. Electrons moving down the electron transport chain
   C. Protons moving down a concentration gradient
   D. Sodium ions moving down a concentration gradient

2. Which of the following processes occurs in both respiration and photosynthesis?

   A. Calvin cycle
   B. Chemiosmosis
   C. Citric acid cycle
   D. Krebs cycle

3. What is the cause of the cramps you feel in your muscles during strenuous exercise?

   A. Lactic acid fermentation
   B. Alcohol fermentation
   C. Chemiosmotic coupling
   D. Too much oxygen delivery to the muscles

4. Which of the following statements is *in*correct?

   A. Glycolysis can occur with or without oxygen.
   B. Glycolysis occurs in the mitochondria.
   C. Glycolysis is the first step of both anaerobic and aerobic respiration.
   D. Glycolysis of one molecule of glucose leads to the production of 2 ATP, 2 NADH, and 2 pyruvate.

For questions 5–8, use the following answer choices:

   A. Krebs cycle
   B. Oxidative phosphorylation
   C. Lactic acid fermentation
   D. Chemiosmosis

5. This reaction occurs in the matrix of the mitochondria and includes $FADH_2$ among its products.

6. This reaction is performed to recycle NAD^+ needed for efficient respiration.

7. This process uses the proton gradient created by the movement of electrons to form ATP.

8. This process includes the reactions that use NADH and $FADH_2$ to produce ATP.

9. Which of the following molecules can give rise to the most ATP?

   A. NADH
   B. $FADH_2$
   C. Pyruvate
   D. Glucose

10. Which of the following is a proper representation of the products of a single glucose molecule after it has completed the Krebs cycle?

    A. 10 ATP, 4 NADH, 2 $FADH_2$
    B. 10 NADH, 4 $FADH_2$, 2 ATP
    C. 10 ATP, 4 $FADH_2$, 2 NADH
    D. 10 NADH, 4 ATP, 2 $FADH_2$